The Agrarian Question in North Vietnam, 1974-1979

A Study of Cooperator Resistance to State Policy

Adam Fforde

An East Gate Book

M. E. Sharpe, Inc.

Armonk, New York   London, England

An East Gate Book

Copyright © 1989 by M. E. Sharpe, Inc.

Available in the United Kingdom and Europe from M. E. Sharpe,
Publishers, 3 Henrietta Street, London WC2E 8LU.

Library of Congress Cataloging-in-Publication Data

Fforde, Adam.
    The agrarian question in North Vietnam, 1974–1979 / by Adam
Fforde.

    Bibliography: p.
    Includes index.
    ISBN 0-87332-486-2
    1. Collectivization of agricultural—Vietnam—History—20th
century. 2. Collective farms—Vietnam—History—20th century. I. Title.
HD1492.V5F47   1989
338.7'63'09597—dc19   88–6726
                                                                        CIP

Printed in the United States of America

Contents

Part IV: The Agrarian Question in North Vietnam

Tables

Preface

This book is the first Western academic work based upon prolonged contact with North Vietnamese society, and the personal circumstances surrounding it require explanation.

I first went to Vietnam in late 1978 to study Vietnamese at Hanoi University and to research my doctoral thesis.[1] I was then twenty-five and believed myself a Marxist socialist. I had been greatly influenced by my time at Birkbeck College, London, where I had done my Master's degree in 1976–77 before going to Cambridge. I left Vietnam after over twelve months' work in late 1979, and I believe that I am still the only Western academic to have studied at Hanoi University for such a long period. I then spent nearly three years writing up, finishing, and submitting the thesis. During that time I acted as a volunteer counselor and interpreter with those Vietnamese "boat-people" who started arriving in the U.K. in late 1979. Most of the Vietnamese I got to know in this way were of rather ordinary social origins and came from the North.

In late 1985 I returned to Hanoi as a holder of an ESRC Post-Doctoral Research Fellowship for a study of Vietnamese industry. I was attached for six months to the National Economics University (previously the Economics and Planning University) as a "Researcher" (*Nghien cuu vien*). To my knowledge I was the first Western researcher to be given that privilege. By that time I no longer called myself either a Marxist or a socialist.

These personal details attempt to illuminate the changing perspective of my work. Vietnam and the Vietnamese mean much for many people in the West; it is better for my own circumstances to be overt. I am in fact rather too young to be a member of the Western generation for whom the Vietnam War has such intense importance and meaning. During the Tet Offensive of 1968 I was fifteen. In any case, I took no active part in the antiwar movement. Many of the Vietnamese academics I have known over the years were officers in the Vietminh or the Vietnamese People's Army. Most Red River delta communes have graveyards set aside for the war dead. Many were bombed or straffed by U.S. airplanes. Yet, though the period covered by this study includes the later stages of the so-called Second Indochina War, the North Vietnamese primary sources upon which it is based were—like myself—not particularly concerned with it. Both they and I were far more interested in the operation of the rural social organizations to which the majority of the North Vietnamese population belonged. This is not to say that life went on regardless, but simply that there was—and is—far more to "Vietnam" than war.

In my own opinion, indeed, much of the Western interest in Vietnam and the Vietnam War has unfortunately been transformed into an inward-looking concern with what it all meant for the West alone. Indeed, given the large numbers of people, especially in the United States, who are linguistically qualified to tackle the field, the limited extent of the available research is rather astonishing. There are to date almost no decent and accessible (i.e., in English) novels or other works of fiction that take as their subject the vivid Vietnamese experiences of past decades. In the end the reasons for this are not hard to fathom, but it goes some way toward explaining why I find the "Vietnam War," as a solely Western preoccupation, uninteresting. Given my feelings for those whose lives were so deeply affected by it, I also find it rather sad.

This study is essentially about what happened when the Marxist-Leninists who governed North Vietnam decided that an inappropriate form of farming—the collective production units called agricultural producer cooperatives—would be the universal norm for North Vietnamese peasants. The inappropriateness of these units meant that there was popular pressure to reduce the collectively farmed area, as well as to restrict other elements of the collective economy. This was the essential nature of the "Agrarian Question" in North Vietnam. Confronted as a result by the very early 1970s with major difficulties in getting the cooperatives to function, and unwilling to abandon the whole project, the North Vietnamese leaders sought instead to strengthen them by giving them greater powers. These were embodied in the so-called New Management System (see chapter 4). Peasant resistance to implementation of this system forms the general theme of this study. It therefore deals, from a somewhat distant academic perspective, with something of great practical importance to the majority of ordinary North Vietnamese. Most of the people involved were—and remain—extremely poor, and the whole exercise led to a not inconsiderable waste of energy and effort. It was probably obvious quite early to many of the people involved that this would be the case, yet it happened anyway.

It is not difficult to argue that the combination of collectivized agriculture and the much-vaunted New Management System was one solution to the considerable developmental problems of North Vietnamese agriculture. Counter-arguments are also easy to envisage (see part IV). At the same time, however, collectivization met certain goals that had little to do with development as such: first, it was consistent with the tenets of Marxism-Leninism and what top leaders had been saying was Revealed Truth for the past twenty years; second, it offered jobs to the local party leadership that were intended to give their social positions both meaning and command over economic resources; third, it sought to bring the rural population's economic activities under a deep and penetrating control. This study shows that the system did not work. In practice, reality refused stubbornly to act in accordance with the tenets of Marxism-Leninism, for the local party leadership could—usually—get along quite happily without it, while the rural population typically succeeded in subverting much of the system while still sending their children to die fighting in the South. The people whose careers suffered most were probably those junior cadres who spent years trying their best, but failing to get the peasants to obey the orders from above. Certainly the aging national leadership largely survived the crisis in social and economic policy that arose in 1979–1982 as various pigeons came home to roost in the wake of the Chinese and Western aid cutoffs.

This is not a comparative study. Although agricultural producer cooperatives and fully collectivized systems are commonplace in Communist countries, rural conditions in North Vietnam appear highly unusual. Collectivization faced a number of difficulties, of which the innate absence of significant scale economies in wet-rice cultivation was perhaps the most important.[2] In addition, the North Vietnamese peasantry had considerable historical experience to draw upon in its dealings with the centralized modern state. The region was particularly poor, and its extremely high population densities were associated with traditional patterns of social organization designed to reduce risk. The widespread popular support for the war against the United States meant that rural collectives became part of wartime propaganda linking the socialist and nationalist movements. At the same time the need to retain popular support mitigated against the use of severe measures to enforce central directives on cooperative management.

This study takes an analytical and developmental perspective that treats the subject distantly and rather coldly, in keeping with the norms of Western academic discourse. It goes into considerable and perhaps somewhat tedious detail in order to argue through various issues on the basis of inadequate data. This permits the discussion to remain academic and analytical rather than contentious and polemical. This should not, though, obscure the basic point, which is that most of the people involved were usually trying as best they could to get around the attempts of the government to make them do things they did not want to do. The official media in Vietnam may have been reiterating the eternal verities of Marxism-Leninism, with minor shifts as political balances altered and objective reality butted in, but in essence much remained deep confusion: life was not like that, and most people knew it—even if they did not always say so. It was only in the early 1980s after the profound "atmosphere change" created by the Sixth Plenum of August 1979 that officially sanctioned public debate started cautiously to recontact reality. The considerable benefits for the economic development of the entire country that resulted are discussed in a planned study of Vietnamese industrial policy, 1979–1985.

The book has a somewhat unconventional layout. A long introductory chapter looks at a number of necessary preliminaries that explain the disciplinary context of the study, the historical and macroeconomic backgrounds to the operation of collectivized agriculture, and the basic issues. This provides the basic framework for the subsequent discussion, which has three main parts. These aim to introduce the reader slowly to what is a detailed and unfamiliar institutional environment, and, as the book progresses, provide pointers to the analytical issues.

A basic theme of the book is the idea that knowledge of formal, or legally constituted, social structures does not necessarily tell us much about underlying "reality." But those structures were nevertheless important because they were the party's "norms," so it follows that to understand what was going on the reader has first to learn what the formal structures were, and then partly repeat the process so as to learn their real meaning. This can be confusing. So, in chapters 2 and 3 the study concentrates on presenting a cooperative's formal structure, while examining the underlying conflicts that occur within collectivized systems. A simple typology is derived from these considerations, which essentially sees the basic difference between cooperative types as being their degree of compliance with party "norms."

The argument then goes in two directions. First, it looks at the immediate origins of official policy and the Agrarian Question of the early 1970s. This then leads on to an examination of the details of the New Management System, which was official policy in the mid-1970s. Experiences in two "model" cooperatives show the "norm" as it was meant to work in practice. Second, it starts to consider how the New Management System would likely operate under normal conditions, given the earlier discussion of cooperative types and associated considerations regarding local socioeconomic conflicts. In part III these converge in a detailed look at the available empirical evidence. This has two stages. First, a highly "nominalized" cooperative is examined in some detail (chapter 6). Second, detailed evidence is gathered from a sample of cooperatives in order to assess the economic consequences at the micro level of the collectivized system (chapters 7 to 10). Finally, in part IV the discussion returns to the level of policy and evaluation. The Agrarian Question of the 1970s is thus understood in terms of policy nonimplementability, with correspondingly important implications for the precise "scientific" nature of Marxism-Leninism. The final chapter discusses some general implications of this.

Editorial Notes

Vietnamese conventionally divide Vietnam into three areas: the North (*mien bac, bac bo*), Center (*mien trung, trung bo*), and South (*mien nam, nam bo*). This corresponds approximately to the colonial French division of the country into Tonkin, Annam, and Cochinchina. Western practice uses North Vietnam to refer to the Democratic Republic of Vietnam (DRV) and South Vietnam to the now defunct Republic of Vietnam. The latter meanings are generally used here. When the Vietnamese sense is intended the context should make that clear.

In Vietnamese, words of more than one syllable are not written continuously (e.g. "cooperative"—*hop tac xa*, is not written "hoptacxa"). Place names such as Hanoi are therefore written Ha noi, and the name of the country itself is written Viet nam. Here the normal Western writing has been kept for those words familiar in the West, such as Vietnam, Hanoi, and Haiphong. Other words are written in the Vietnamese manner. Diacritical marks are omitted from Vietnamese words throughout the text, with the exception of the Glossary.

Acknowledgments

My first acknowledgment is to the late Suzy Paine, who taught me development economics at Birkbeck and supervised my doctoral thesis. She gave me my head at a time when her own views were changing rapidly, so long as I could defend myself against her periodic critical onslaughts. She is sorely missed. I also owe much to the staff of the Vietnamese Language Faculty of Hanoi University, and to various members of the Political Economy and History Faculties, who gave their time to helping me learn something about their country. Like Suzy Paine, they would find much to criticize in what I have written, but that is in the nature of academic study.[3] I also owe a particular debt to James C. Scott, who, by inviting me to a Conference on Everyday Forms of Peasant Resistance,[4] encouraged me to start opening up my thinking and reinterpret policy nonimplementation in terms of cooperator resistance to state policy. Many people both within and outside the small freemasonry of Vietnam scholars have been kind enough to read and comment upon this work in previous drafts. I would like to thank in particular the following for their comments and criticisms: David Marr, W. J. MacPherson, Alexander Woodside, John Kleinen, Samuel Popkin, Karl-Eugene Wadekin, Andy Vickerman, Melanie Beresford, and an anonymous publisher's referee. My two thesis examiners, Christine White and Peter Nolan, provided many useful comments and suggestions. Teresa Halik of Warsaw University helped me to a better understanding of Vietnam and provided valuable and comradely assistance to me during my stay in Vietnam.

I would like to thank John Wells and W. J. MacPherson for their assistance and support at critical moments during my time at Cambridge. In addition, my thanks go also to Jenny Haselden, who typed various drafts with great tolerance for the quality of the copy and the problems involved in interpreting my handwrit-ten Vietnamese. I am grateful to Jackie Allen and Jan O'Brien for helping transcribe text into the word processor. I would also like to express my apprecia-tion to my friends in Landbeach village for their understanding and sympathy, especially to Jimmy and Clare Tait, Mrs. P. J. Patterson, and Jack Ravensdale, and also to my friend Rick Lewis.

This study has been primarily funded by the Economic and Social Research Council under their Research Studentship Scheme. Funds for study in Hanoi were supplied at short notice, for which I am grateful. In its closing stages the work was assisted by loans from the National Westminster Bank and Clare College, Cambridge. I am responsible for the contents of this book.

Abbreviations

*	Words in text marked with an asterisk are standardized translations of technical specialized Vietnamese terms—see Glossary.
A	Active cooperatives of the sample.
ASM	Active but strongly managed cooperatives of the sample.
AWM	Active but weakly managed cooperatives of the sample.
CB	*Cong bao* (Gazette)—the official Gazette of the Hanoi government, whether the DRV or SRV.
DRV	Democratic Republic of Vietnam ("North Vietnam").
Hanoi I, II	Nonnumeric and numeric data collected at Hanoi University 1978-79.
KTVH	The collection of official North Vietnamese data—*Su phat trien kinh te va van hoa mien bac xa hoi chu nghia Viet nam* (The Cultural and Economic Development of Socialist North Vietnam), Hanoi, 1978 (in Fforde and Paine, "The Limits of National Liberation," London, 1987).
M	Model cooperatives of the sample.
NCKT	*Nghien cuu kinh te* (Economic Research)—the journal of the Economics Institute, Hanoi.
NCLS	*Nqhien cuu lich su* (Historical Research)—the journal of the Institute of Historical Research, Hanoi.
NMS	The New Management System prescribed for agricultural cooperatives' management structures during the mid-1970s.
SRV	Socialist Republic of Vietnam (post-1976 reunification Vietnam).
SLTK	*So lieu thong ke Cong hoa xa hoi chu nqhia Viet nam 1981* (Statistics of the Socialist Republic of Vietnam 1981), Hanoi, 1981.
UN	Written source supplied by the University of Hanoi.

Chronology

939 Recovery of independence after more than 1,000 years of Chinese rule.

11th and 12th centuries under Ly dynasty (1009–1225) Buddhist monks gradually lose positions of influence at court, replaced by Confucian scholars.

1407 Chinese occupation of country: "kulturkampf" destroys many historical sources.

1428 Accession of Le Loi, founder of Le dynasty (1428–1786), who implements a state-regulated system of land distribution. Systematic attempts to use Confucian-based Chinese administrative system.

1527 First period of national division: Nguyen lords in South, Mac 1592 imperial interregnum in North.

1592 Mac defeated by alliance between Trinh and Nguyen families. Second period of national division—intermittent civil wars continue; Le emperors rule in name only in North, dominated by Trinh.

1774 Tay-Son revolt—after rapidly acquiring popular support, movement deposes Le (1786) and defeats invading Chinese army (1789).

1802 Final defeat of Tay-Son by member of Nguyen family with Western assistance; takes name Gia-Long to found Nguyen dynasty (1802–1945).

1858 Tourane expedition heralds imminent French conquest of South Vietnam.

1862 Treaty cedes three provinces around Saigon to France.

1874 Treaty confirms French title to Cochinchina.

1879 Civilian government established in Cochinchina followed by rapid development of Mekong delta.

1882 Hanoi seized by large French expedition.

1883 Hue captured.

1884 China recognizes Franco-Vietnamese treaties, ending its claim to suzerainty.

1885–1887 French consolidate their rule in Tonkin and Annam, setting up puppet emperor and establishing Indochinese Union of Vietnam, Cambodia, and Laos (1887).

1897–1902 Paul Doumer (governor-general) establishes centralized control over Indochina from his capital in Hanoi. Colonial state apparatus consolidated.

1902 Hanoi University founded.

1915 Last Confucian exams held in Tonkin—abolished in Annam in 1918.

1929 Founding of Indochinese Communist Party in Hong Kong.

1930 Rising by Vietnamese Nationalist Party (Viet Nam Quoc Dan Dang, founded 1927–28) leads to capture of its leadership.

1930–1931 Nghe An and other peasant risings spurred by effects of depression lead to violent French reprisals and near elimination of Communist leadership.

1940 Japanese occupy Indochine but leave French in control of administration. General rising of Communists in South put down and party network almost obliterated.

1941 Founding of Viet Minh and establishment of guerrilla bases in Northern highlands.

1942 Nationalist Chinese finance both Communists and Nationalists in Vietnam. Partial reestablishment of Communist network.

1944–45 Famine in North Vietnam kills approximately one million.

1945 Japanese disarm French, creating power vacuum (March). Emperor Bao Dai abdicates (August). Declaration of Independence in Hanoi by Ho Chi Minh (September). Creation of Democratic Republic of Vietnam. Chinese Nationalists move into North to accept Japanese surrender. British troops in South accept French rearmament and use Japanese soldiers to crush Viet Minh. Disbandment of Indochinese Communist Party (November)—replaced by "study group" within Viet Minh.

1946 Preliminary Franco-Vietnamese agreement (March). Haiphong shelled by French ships after breakdown of talks (November)—Viet Minh retreat to mountains.

1949 Victory of Chinese Communists; Canton taken in October.

1950 French forced out of border areas allowing Viet Minh easy access to Chinese matériel. DRV recognised by Communist bloc (January).

1951 Inauguration of Vietnam Workers' Party, founding of party daily *Nhan dan* (February). Promulgation of unified agricultural tax, founding of National Bank (May).

1953 Fourth Plenum of Politburo (January) passes provisional outline of party's land reform policy. French set up advance base at Dien Bien Phu (autumn). Land reform law passed (December).

1954 Fall of Dien Bien Phu (May) on eve of Geneva Conference—French agree to leave North Vietnam (July). Early stages of land reform. First meeting of Central Land Reform Committee (June).

1955 Final departure of French from North Vietnam—country divided at 17th parallel. Elections prior to reunification never held. Formalization of land reform (March): "people's tribunals" replace "peasant struggle meetings." Land reform extended to newly liberated areas.

1956 Official end of land reform (August): Ho Chi Minh publicly admits to errors. Tenth Plenum (September) admits to land reform errors. Resignation of Truong Chinh (party general secretary) in wake of land reform (November).

1956–57 Nhan Van affair parallels Hundred Flowers in China. Repression of intellectuals critical of the new government ends with party hegemony over public intellectual expression. Correction of Errors in countryside sees pre-land reform local leadership partly recover its position.

1957 Twelfth Plenum sets goal of ending reconstruction and moving on to planned development.

1958 Fourteenth Plenum examines Three-Year Plan (1958–1960).

1959 Monetary reform (1:1,000). Fifteenth Plenum formally decides upon return to armed struggle in the South. Sixteenth Plenum and speech by Truong Chinh to National Assembly herald rapid implementation of agricultural cooperativization.

1960 Movement to establish joint state-private enterprises basically complete (August), with 97 percent of both family units and capital. Third Congress (September) approves First Five-Year Plan (1961–65). Movement to establish lower-level cooperatives in agriculture basically completed. Movement to set up lower-level artisanal cooperatives basically completed (October), with approximately 80 percent of family units.

1961 Incoming U.S. President Kennedy told by Eisenhower that Laos (not South Vietnam) is "the key to Southeast Asia" (January). U.S. troop numbers in South Vietnam exceed limits agreed at Geneva (April). Fifth Plenum (July) affirms "reforming and improving the management of cooperatives is the most important element in developing [agricultural] production."

1963 Politburo (March) issues directives on cooperatives for the period 1963–65, focusing upon management methods and the role of the party. Eighth Plenum (April) discusses progress of Five-Year Plan. Rising macroeconomic tensions resulting from overambitious development targets manifest in growing import dependency and domestic inflationary pressures. Beginning of stage 1 of campaign to "improve cooperative management" (June). Great Buddhist opposition to Ngo Dinh Diem in South (October); Diem assassinated November 1. Kennedy assassinated November 22. Vice-President Johnson takes office pledged to continue U.S. commitment.

1964 Tonkin Gulf incident (August) allows U.S. administration to pass legislation allowing overt use of force in Vietnam.

1965 U.S. commences "Rolling Thunder" bombings of North Vietnam. Crisis measures taken by Hanoi include decentralization of state organs and industry and general mobilization. Beginning (October) of stage 2 of campaign to "improve cooperative management."

1966 Heavy U.S. bombing in attempt to prevent continuing support from Hanoi for guerrilla and main-force units in South Vietnam. Culminates in the events of 1968: "Tet" offensive (January, showing apparent strength of antigovernment forces in the South), refusal of President Johnson to stand for a second term (because of domestic opposition to the war), and announcement of "unconditional" bombing halt (March) and first meeting in Paris of U.S. and Vietnamese negotiators (May).

1968 Shift from lower-level to higher-level agricultural cooperatives basically finished. Reported staples availability from domestic production falls below 13 kg milled rice equivalent per month for first time. Soviet food aid rises sharply. So-called III contracts controversy between Le Duan (first secretary) and Truong Chinh over validity of household-based farming apparently resolved in favor of policies emphasizing the collective. August decree expands state management of trade, reinforces regulations against speculation and theft of state property.

1969 Death of Ho Chi Minh (September). Report of Central Agricultural Committee (July) reveals declining collectively farmed area in many Northern provinces.

1969–1971 Extensive diplomatic activity. Nixon doctrine allows for steady withdrawal of U.S. troops from South Vietnam.

1971 Nineteenth Plenum (October) advocates reinforcement of local administrative and party organizations in face of difficulties with Northern agriculture. Beginnings of experiments with centralized and detailed methods of cooperative management that became the so-called New Management System.

1972 Visit of President Nixon to Peking (February). Escalation of activities in South Vietnam by North Vietnamese Army (March). U.S. responds by heavy bombing of Hanoi/Haiphong (April) and harbor mining (May). Central Committee makes general appeal to population. Twentieth Plenum (April) advocates need for economy to "advance from small-scale to large-scale-production" and attacks "dispersed small-scale and artisanal methods." Presidential decree on maintenance of law and order and management of trade in wartime (May). "All-out" bombing attacks on Hanoi and Haiphong ordered (December) to accelerate North Vietnamese agreement to compromise.

1973 Paris agreements initialed (January); U.S. bombing finally to cease. Census of state property (October).

1974 National Census. Government Council decides to "abolish the free market in staples"—decision inoperative. Twenty-second Plenum (April) reasserts party hostility to small-scale production and confirms commitment to centralized management methods in agricultural cooperatives. Thai Binh Conference (August) announces "Toward a Large-Scale Socialist Agriculture" and affirms support for New Management System.

1975 Retreat from Central Highlands by South Vietnamese Army (March) heralds rapid collapse of South. Embroiled in aftermath of Watergate, U.S. refuses to provide increased military support. Saigon falls (May).

1976 Fourth Congress of Vietnam Workers' Party. Name of party changed to Vietnamese Communist Party. Formal reunification of country under name of Socialist Republic of Vietnam. Twenty-fourth Plenum analyzes situation in South, criticizes it for import dependency and predominantly small-scale production. Decree 61-CP (April)—"Toward Large-Scale Socialist Agricultural Production"—reaffirms official commitment to New Management

System. Plans laid for extension of North's collectivized agrarian system to the South and Center.

1977–78 Growing problems with the economy; bad weather and collectivization hit Mekong delta. Procurement levels fall. Drive against private sector in South. International relations deteriorate, leading to Vietnamese "involvement" in Cambodia (late 1978) and membership in Comecon. Chinese and Western aid cuts sharply reduce share of available resources controlled by state. Second Plenum (1977) confirms commitment to Thai Binh Conference line on cooperative management—strong support for collective pig herds. Decree 55-CP (March 1978) encourages shift to ration-based distribution in some Northern cooperatives as collective incomes fall.

1979 Chinese "punitive" action in Vietnam's Northern provinces. Collectivization of Central provinces basically finished; collectivization drive in Mekong temporarily abandoned. Procurement reduced to crisis levels. Many Northern cooperatives under great pressure from members to reduce still further the extent of collective production in favor of own-account activities, but Fifth Plenum (February) holds to old line, advocating extension of centralized methods used by model cooperative Vu Thang. But Sixth Plenum (August) heralds sharp change in "policy atmosphere" without changing cooperative management policy—concentrates on industry. Decrees late in year allow cooperatives to give out "idle land" for farming by individual cooperators, and process of "decollectivization" gathers pace in the North. Decree 373-CP reinforces state's monopoly right to trade in staples and other goods.

1980 Agrarian policy subject to great debate; underlying trend toward controlling decollectivization and reestablishing some sort of central control over the collectivized system. Some recovery in procurement and staples output, mainly in the Mekong, aided by priority in state supplies. Output contract system starts to surface as new policy, aiding recollectivisation in some areas. Ninth Plenum (December) heralds "new policy" and calls for "expansion of the implementation and improvement of forms of contracting in agriculture."

1981 Haiphong Conference (January) and Party Secretariat Order 100-CT/TW introduce output contracting as, inter alia, a way of "strengthening socialist relations in the agricultural cooperatives" and improving cooperative management. Situation now relatively stable and crisis over.

The Agrarian Question in North Vietnam, 1974-1979

1

Introduction

Context of the Study

This study argues that the so-called New Management System of the mid-1970s was essentially an attempt to reinforce the managerial structures of the agricultural producer cooperatives that were the intended vehicle of socialist transformation in North Vietnam. By that time, it was clear that there was a widespread and spontaneous tendency for the cooperatives to become "nominalized" and therefore far less important and powerful than intended. Many peasant-cooperators were deeply dissatisfied with the operation of the fully collectivized agrarian system. This confronted the Vietnamese Communist leadership with an evident and profound failure in their overall strategy. As Marxist-Leninists, they had argued that the combination of National Liberation and socialist forms of social organization were both inseparable and historically correct: they were the only logical and moral course possible for Vietnam and the Vietnamese. Thus the nonimplementability of their rural policies had immense and destructive implications for their arguments, and therefore for their own positions. And, as will be seen, they reaffirmed the correctness of these policies after national reunification in 1975, as they attempted to extend agricultural collectives and the other elements of the socialist revolution to the South.

While such profoundly political considerations form the basis of the Agrarian Question in North Vietnam, the foreign observer and analyst has to establish his or her own disciplinary position and analytical framework. This is the first main aspect of the overall context of the study.

3

Disciplinary Perspectives

North Vietnamese collectivized agriculture can be studied from a number of different perspectives. The comparative Communist systems approach, development studies, and Sinology all have much to contribute to a better understanding of the operation of collective farms in North Vietnam.

As a discipline, development studies tends to focus upon processes of socio-economic change and the interaction between changes in social organization and economic growth. In North Vietnam the collective farm was presented as part of the socialist revolution, and the basic "developmentalist" element of Vietnamese Communism should not be forgotten (see chapter 4). Unlike French colonialists, the postindependence rulers were strongly committed to attaining rapid and independent economic growth. Marxist-Leninist theory presented the collective farm as one part of the institutional basis for that growth. Yet development studies, with its predominantly Western (or at least non-Communist) experiences, often suffers from a tendency to confuse form with content when confronted with Communist cooperatives. Unlike, say, the members of a single producer cooperative in an Indian village, peasants belonging to the collectives of a fully collectivized Communist system have little real choice over whether to remain in the collective or not. Once the collective system is established, social and economic sanctions against nonmembership are usually so severe that attention inevitably focuses on how to make the collectivized system work, rather than on how to leave it. But while the central authorities determine the form of rural institutions, local practice may be free to determine their content.

From a development studies perspective it is above all the highly centralized nature of the Communist system that makes it unusual. For many reasons, this means that changes in prescribed institutional forms result from decisions taken centrally rather than locally. Furthermore, the initial "revolutionary" changes extend extremely deeply into rural society. Peasants are confronted with a package of centrally formulated policies that call for fundamental alterations in their way of life. What are they to do? Resistance to unwanted elements of the package can take two qualitatively different paths: changing the policies and resisting their implementation. Thus in a Communist collectivized agriculture the key processes whereby social organizations adapt to changing circumstances during development (their "endogeneity") occur in two quite separate ways: first, through local responses to the day-to-day practicalities of implementing national policy, which determine the effective content of collective forms under the particular conditions confronting them; second, through the response—if any— of the central policy-makers to agrarian problems.

This study deliberately focuses upon the former and examines grass-roots North Vietnamese experiences with a system of management prescribed for the rural collectives (the New Management System) that was in practice hard to implement. This brings out the central authorities' inability to determine the content of local institutions. The study shows how the policy response to this Agrarian Question basically sought to extend central influence over rural affairs. This deep-rooted drive for central control is a characteristic of neo-Stalinist government. From a developmental perspective, therefore, North Vietnamese experiences are, while not unique, rather unusual.

From a comparative Communist systems perspective, however, Vietnamese experiences are also far from ordinary. For a number of reasons discussed at length below, North Vietnamese peasants very often successfully avoided implementing undesired central directives. This resulted in an unusually high degree of systematic noncompliance with party directives on the part of rural cadres. Usual assumptions about the internal discipline of a Marxist-Leninist party do not seem to hold in the Vietnamese case—at least not in the area of rural economic management. Yet the basic model governing the form of rural social organization in North Vietnam—the agricultural producer cooperative—came directly from Soviet experience during the 1930s.

All ruling Communist parties have at some time sought to establish systems of collectivized agriculture. As the first Marxist-Leninist party to attain state power, the Communist Party of the Soviet Union presided over the events of 1929-1932, out of which came the *kolkhoz* form, based upon the *artel*, which provided a dominant model for subsequent collectivization drives in other areas.[1] A major variant of this system was the now abandoned commune-based strategy of the Chinese Communist Party, with its strong emphasis upon the integration of various hierarchically organized teams and brigades into very large units (the communes) that had both administrative and economic responsibilities.[2] In practice, experience has shown the great variety of possible outcomes—compare, for instance, the apparent success of Hungarian producer cooperatives with the chronic and expensive inability of Soviet agriculture to meet plan targets.[3]

As Sinologists persistently tell us, there is much scope for comparative studies of Vietnam and China. The basic Vietnamese technical vocabulary is indeed Chinese in origin (see glossary), and surface similarities are intriguing: for instance, the drive toward increasing scale and a coincidence of administrative and economic units of management, and then the shift toward some form of contracting with constituent households from around 1979-1980. But the Chinese themselves drew upon the Soviet model when they collectivized in the 1950s, as did the Vietnamese. Furthermore, it appears that the weight of the collectivized system was far greater in China, where—unlike North Vietnam—cooperators' rights to free market output disposal were for periods effectively curbed in some areas. The lower level of preexisting economic surplus in North Vietnam meant that inflationary pressures during the early stages of development were greater, which, coupled with wartime inflation, shifted the balance against collective as opposed to own-account economic activity. To these factors might be added others: reduced procurement pressure in North Vietnam because of food aid imports; a more laissez-faire attitude, especially toward use of the army, perhaps because of the need to retain popular support during wartime; the Vietnamese psychology, with its tendency toward pragmatic compromise and lack of blood-letting among the top leadership. But this list of factors, while thought-provoking, also reveals the inevitable tendency toward premature generalization in Sino-Vietnamese comparative study at present, when "reform" is so much in the air. This is true both for Chinese and for Vietnamese studies. As centralized systems, it is also in the end hard to see how one can reasonably compare an area with a population of around twenty million (the DRV in the early 1970s) with a country the size of China.

Western research into the socioeconomic problems of Communist-ruled areas of Vietnam remains in its infancy. In preparing this research, the present writer was surprised to find that many crucial background areas had almost to be approached from scratch: we lack, for example, a proper and complete economic study of the Vietminh "liberated areas" during the period 1946-1954, and, despite the existence of comparatively abundant vernacular source materials, no detailed analysis has used them to study the evolution of economic policies during the First Five-Year Plan (1961-1965).[4] Two crucial sets of problems are those of the wider historical background to the collectivization drive of 1959-1960 and the macroeconomic environment within which cooperatives had to operate.[5] These are discussed below.

Study of the Political Economy of Rural North Vietnam

The basic set of questions that this study asks are those of orthodox political economy: what were the socioeconomic factors that determined the processes of production, consumption, distribution, and exchange? The work is particularly concerned with those factors influencing the level and efficiency of resource mobilization. This is because of the importance attached by political economy to a proper understanding of the effects of social institutions upon economic growth and accumulation. Basic to such an investigation is an analysis of the conditions of labor's access to means of production, of the social control of that production, and of the appropriation and distribution of the resulting output. Material and other incentives acted both within the formal organizational framework and, via institutional change, upon it.

The empirical basis of this study is weak, but in the Vietnamese context this is unfortunately inevitable. Based upon a number of North Vietnamese microstudies from the period (both published and unpublished) and some very limited field data gathered in 1978-79, it attempts to show how the collectivized agriculture of North Vietnam was frequently the scene of sharp social and economic conflicts. It has been estimated that around 75 percent of cooperatives (see chapter 4) had only a nominal existence, and a cooperative's full statutory rights to control property, production, and output distribution were rarely fully recognized. This does not mean that the cooperatives simply became ghosts, however, for collective production continued under brigade management. It was the nominalization of the central organs of the cooperative, through which higher levels sought to control it, that was the problem. Managerial reform had frequently to take the form of a struggle to impose effective central control over collective assets, enforce labor norms, and so forth.

The Immediate Historical Context

The study focuses upon experiences with the set of organizational principles for cooperatives explicitly advocated during a comparatively short period—1974–1979. The North Vietnamese peasantry had been collectivized rapidly in 1959–1960 (see chapter 4). By the time of the 1973 Paris Agreements the party was moving toward solving the existing Agrarian Question by further strengthening the organization of the agricultural producer cooperatives to which 96 percent of peasant households, with 95 percent of the cultivated area, then belonged.[6] Although policy debates and experiments with the New Management System (NMS) go back to at least the turn of the decade, the important Thai Binh Conference of August 1974 saw the leadership endorse the detailed set of policies known as the NMS. This system then became the prescribed norm by which party orthodoxy judged cooperative performance. But the NMS represented a further extension of well-entrenched and strongly centralizing views dominant among members of the Vietnamese leadership and was a logical extension of previous policies.

The continuing commitment to centralized management methods was quite in keeping with contemporary thinking in the Soviet Union and other members of the CMEA (Comecon). Confronted with major problems in increasing food supplies as consumer demand rose in line with higher real incomes, policy makers in CMEA countries tended to stress the following concrete measures:

1. Accelerated supplies of industrial inputs to agriculture, greater fixed capital investment in agriculture, and development of nonagricultural lines of production by agricultural collectives.

2. Comprehensive development of "agro-industrial complexes."

3. Improved organization and productivity of labor.

4. Cooperation and concentration in agricultural production.

5. Development of socialist international agricultural integration.

6. An overall drive to reduce the CMEA's perceived aggregate shortage of supplies of food to consumers and agricultural inputs to industry.[7]

All of these points were part of orthodox Vietnamese Communist thinking at the time. It is important to note that this overall package was quite contradictory to the idea that a decentralization of cooperative management was a prerequisite for improved economic performance. But such a decentralization was the direction in which local interests were pushing throughout the period. It was also the direction in which policy eventually moved after the 1978–79 crisis. In this context it should be noted that there is evidence that the Vietnamese leadership had decided by 1974 to move toward the Soviet Union and away from China.[8] Although this study cannot analyze in detail the relations between Vietnam and other Communist countries, the above does strongly suggest that there was nothing in Vietnamese policy with which the Soviet Union could at that stage take great issue.[9]

It is likely that the leadership did not expect to find themselves confronted in 1975 with the task of managing a reunified country. The crucial new element that this introduced into the economic policy equation was the Mekong Delta and its possible socialist transformation (i.e., collectivization). The delta's large potential food surplus transformed the underlying economic situation by apparently removing a major constraint on development—inadequate supplies of consumer goods. In the event the leadership did not change course. In 1976 they reaffirmed the NMS as correct policy for the North and moved toward collectivization of the Mekong Delta and the Center of the country.[10] The NMS was therefore an integral part of the package of essentially neo-Stalinist and highly ambitious policies that the leadership sought to implement in the immediate postreunification period. From this perspective its continuity with earlier policies is important, and this relates also to its role in uniting ideologically the nationalist and socialist revolutions both before and after national reunification (see chapter 4).

The late 1970s were characterized by a gathering economic crisis based upon fundamental flaws in the overall development strategy. These economic problems were given an added and far greater impetus by the Chinese and Western aid cuts of 1978–79. In the event, the Sixth Plenum of August 1979 saw a substantial reversal in the "policy atmosphere" that led in turn to a rapid change in the official line, so that by late 1980 the NMS was effectively condemned (see chapter 12). Its highly centralized management methods were replaced by a decentralized system of "output contracting" (*khoan san pham*) between the cooperative and its constituent households that entailed a radical change in the basic norms of cooperative organization.[11] From this perspective the NMS's political role and importance start to become clearer: it was a policy that failed and had to be abandoned, and it was a policy that the old nationalist/Communist leadership had fully approved. Such issues are discussed further toward the end of the book, when the grass-roots problems the NMS confronted—and therefore the precise nature of the Agrarian Question—should be clearer to the reader.

The Historical Background

Traditional Rural Organization

Although parallels should not be overstretched, Vietnamese historical experiences create striking echoes in contemporary rural society, most especially in the opposition between centralized state and relatively autonomous rural periphery. The much-disputed "reunification" of the country under the Tay Son (1786) was followed by the establishment of the Nguyen dynasty (1802), who ruled Vietnam from Hue in the Center. Until the French conquest, Vietnam was then ruled by a Confucian court that sought to implement its own interpretation of Chinese (Qing) orthodoxy.[12] In this it followed long tradition during a period of conservatism and reaction. Even then, however, it was recognized that the indigenous Vietnamese system of rural organization was quite satisfactory and should be defended, despite its lack of congruence with the maxims of "Northern scholars."[13]

Unlike many South Chinese rural communities, Vietnamese peasants deemphasized extensive kinship ties. They instead tended to replace clans by administrative organizations to which (subject to certain limitations) individuals could belong independently of their ancestry.[14] The state correspondingly recognized an independent administrative level—the commune, or *xa*—as the basic unit with which it would deal. The Vietnamese commune was, for the peasantry, the point at which the state's demands for taxes, corvee, and military labor were felt; corporate responsibility for meeting such demands helped reinforce the characteristic antipathy toward excessive private socioeconomic power. In a highly risky climatic environment collective social organization had great value. Since the commune therefore met the basic demands of both the state and much of the peasantry it is not surprising that it was seen as an institution of great importance.

Two key elements of the commune were its communal lands and the commune-based religious and ritual activities. The former frequently occupied well over 25 percent of the cultivated area. Farmers apparently cultivated them as wet-rice paddy in a way similar to the private lands. In principle, communal land was let or allocated out in a fair distribution to meet what were seen as essentially communal ends, taking account of the need for income support among families with poor labor endowments. Rent was used to meet communal expenses, such as the upkeep of temples and other buildings. The responsible agency was the council of notables (*hoi dong ky muc*), which was thereby usually the largest landholder in the commune. The dynastic state viewed this council as the agent responsible for the commune. The notables, who were not necessarily large private landowners, had high status and derived much of their personal authority from membership on the council. Membership was attained through some form of election, which often recognized preexisting status—for example, that of the retired mandarin.[15] Collective authority was further enhanced by the council's power to allocate the state's demands (i.e., for taxes, corvee, and military service) among the commune's households. Although the state in principle regulated the way in which this was done, the commune's independence meant that local interests had the determining role.

Commune-based religious and ritual activities provided a focus for the ebb and flow of family fortunes and alliances that dominated local society. The commune's tutelary spirit was often the gift of the emperor and had nationalist significance. It was housed in the commune's long-house or *dinh*. These buildings, many of them in considerable (if subdued) taste, also provided fora for meetings of the council of notables and feasts recognizing formal status rankings. Such activities helped define the commune as the basic unit of rural life. In addition, they also reinforced two crucial ways of thinking. First, people tended to believe that local status was largely to be achieved through the attainment of positions within formally constituted bodies. In the dynastic system the most important was of course the council of notables. Second, a key role of rural collective organization was that it should act as an interface between local interests and higher authority. The commune was a protection against the state. Both of these ideas have strong echoes in rural affairs during the 1970s. Collective institutions were able to satisfy both the centralized state's needs for resources and many peasants' desire for insurance against risk. They could then become the natural focus for social competition both within the local collectivity and between it and higher levels of the state hierarchy.[16]

French Colonialism and the Nationalist Vietminh

During the colonial period rapid demographic growth and abrupt changes in the environment within which the communes operated saw a sharp increase in both the risk and reality of land loss and debt-bondage for ordinary peasants. Notables and others used the now favorable attitude of the state toward land concentration (and especially toward mortgage foreclosure) to accumulate private wealth. In this the sharp increase in the demand for cash brought on by the colonial state's higher and now monetized tax burden helped them considerably. These new styles of social advance no longer involved the pursuit of higher positions in the commune's social hierarchy, but instead the use of techniques of usury and the manipulation of commodity markets to amass property and wealth. The poorer peasant, forced by the risky environment to borrow for consumption or to pay taxes, saw a previously tolerable patron-client relationship turned into a usurous burden that gradually separated him from his traditional property rights. Private land was taken as security for debts, and the communal lands fell increasingly under the control of those notables who were actively building up their personal positions. The consequent loss of minimum income support was one element in the pattern of revolt seen in the central provinces during the onset of the Great Depression.[17] Thus the reduced importance of corporate as opposed to private economic power was a major feature of the colonial period. In a low-income and highly risky environment this process imposed increasingly severe costs upon ordinary peasants in a situation where alternative employment was extremely limited.

The arrival of Vietminh cadres in the "liberated" areas of North and Central Vietnam during the late 1940s abruptly checked this process. Quite apart from the radical-socialist sentiments of the movement, which were in any case somewhat limited in its earlier stages, the tax burden was now payable in kind. In addition, notables could no longer use their contacts in the colonial administration to foreclose on mortgages and manipulate the legal system. Controls on markets may also have been important. It is noteworthy that, if communal lands are included, the redistribution of land in the period up to the beginning of Land Reform proper around 1953 was actually greater than that during the Land Reform campaign itself.[18] And, although this area has not yet been well researched, it seems plausible that the Vietminh's Resistance Administration Committees often acted as a new focus for communal corporate economic activities in place of the largely corrupted councils of notables. Not only were taxes in kind and labor duties still corporate responsibilities, but land taken over from "reactionary" landlords was not allocated to poorer peasants as "private" land. Instead, the commune retained higher property rights and a reallocation would, in principle, take place after five years.[19] This behavior appears quite natural in the light of the traditional attitude toward the commune's role as a distributor of communal property to those "in need." It may also be of interest to note that the DRV did not nationalize land along Soviet lines, suggesting again a certain respect for local property rights.[20]

Land Reform and Cooperativization

Land Reform, from 1953 to 1956, led to major upheavals in rural society. The resignation of party First Secretary Truong Chinh in 1956 demonstrates the severity with which the party viewed the shortcomings of the Land Reform campaign. These problems were associated both with the campaign's assumptions about rural society and with the way in which Land Reform was carried out in practice. Its underlying theory was based upon a Leninist view of the relationship between social and economic power which stressed the importance of landlordism and other forms of "private" economic power. In a highly corporate system such a view was clearly inadequate and therefore inappropriate.

Whereas the earlier "partial" reforms prior to Land Reform had used existing local institutions, the Land Reform campaign based itself upon a separate organization set up parallel to the established hierarchies and answerable to the Central Land Reform Committee. This could often result in sharp local conflicts as people with positions or contacts in the Vietminh found themselves denounced as landlords or rich peasants while their property came under attack. But although Land Reform and the earlier measures effectively destroyed much of the economic base of the landlord-usurer group, they could not end many economic inequalities.

Continued differentiation derived either from access to corporate economic power via membership in the various organizations of the commune or from personal or household attributes such as family labor resources, educational and other skills, and economic resources untouched by the redistributions of land. The local communal leadership in the late 1950s thus tended to divide into two groups. First, there were the more arriviste poor peasant cadres whom the Land Reform teams had chosen for their revolutionary purity; second, and often in opposition to them, were the older-established elements, some of whom had only recently recovered their positions after the Correction of Errors campaign (1956–57).[21]

Cooperativization and Amalgamation

The cooperativization of the North Vietnamese peasantry therefore began in 1959 with little guarantee that the DRV and local organizational structures were sufficiently united to ensure a disciplined implementation of central directives. The initial stages of collectivization in 1959–1960 were extremely rapid. In under two years nearly 90 percent of peasant families (ho) with around 70 percent of the cultivated area joined cooperatives. Of these, however, only around 20 percent were of the higher-level form where there was, in principle, no remuneration for property brought into the cooperative.[22]

Through the 1960s the dominant statistical trend was for a transformation of these lower-level into higher-level cooperatives. This was coupled with increases in their average size in terms of both membership and area so that by 1968 over 80 percent of cooperatives were of the higher-level type.[23]

This period has not yet been properly studied. Although it is true that the landholdings of the new entrants tended to be above average on a per family basis, the premature conclusion that this was therefore a process by which "rich peasants" took over the cooperatives is unjustified.[24] In the first case, this was a period in which collectivization was being extended to the midlands and highlands. In these areas average landholdings were far higher, since cultivators tended to use methods of farming, such as slash-and-burn, that used more land than the intensive rice-cultivation of the deltas. Second, one might as equally argue that "poor peasants" were forcing "rich peasants" to join cooperatives; the former, if well entrenched in crucial positions of authority, were in a far from weak position in any political struggle.

The underlying economic and other interests behind the amalgamations and shifts to higher-level status of the mid- and late 1960s deserve fuller study. To give just one example, it could apparently be profitable for the nonworking parents of cooperators to stay outside the cooperative, with their land. This was because the land would be worked by their children rather than by the cooperative, so there would be no perceived "labor charge." Thus most of those still outside cooperatives in the early 1970s were reportedly old people.[25] This interesting area requires further research before conclusions can be drawn.

By the early 1970s, then, the collectivized peasantry of the Northern deltas had had nearly two decades of experience with Communist rule, first with the

Land Reform campaign, and then with cooperativization and the push to amalgamate cooperatives. They had seen the First Five-Year Plan (1961–65) come and go and, from 1965, they had suffered three years of severe U.S. bombing coupled with the hypernationalism accompanying wartime mobilization. Chapter 2 presents details of the collectivized system. First, however, it is necessary to discuss the economic environment within which the cooperatives operated.

Aggregate Economic Development

The macroeconomic development of the DRV presents a number of problems, not least of which is the relative absence of reliable data. Fforde and Paine (1987) provides an analysis of the economic development of the DRV. The overall development pattern stemmed at root from the neo-Stalinist attempt to implement an unsuitable policy of rapid industrialization in an area of low economic surplus. This resulted in severe macroeconomic tensions, most especially domestically, in the creation of inflationary pressures, and internationally, with the move toward import dependency.[26]

These processes may be sketched as follows. In the early 1960s rapid growth of employment in the state sector generated a demand for consumption goods that was in principle to be met by increased output from the state sector, whether to be exchanged for agricultural products produced in the largely collectivized agricultural sector or by direct exchange with state employees. Aid or the private sector would, in principle, have to bridge any gap that arose. In practice, the increase in effective demand that resulted from the rapid increase in state employment instead facilitated an inflation that led to a large increase in "outside" activities associated with the free market. As prices rose on the free market individuals adjusted their economic activity to take advantage of more profitable opportunities.

As the "inside"—socialist—sectors relied more and more on administrative measures to keep control over economic resources, economic calculation of costs became less important for those who received such goods. Delivery at "low" state prices—that is, at prices well below those on the free market—meant that goods had to receive large effective subsidies, either from aid donors or from the state budget. An aggravated shortage economy therefore became well-established, with economic agents balancing off the costs and benefits of continued participation in the administered and supply-constrained socialist sectors with the high prices and cash incomes available on the free market. This resulted in considerable static economic inefficiency, in the sense that a more economically efficient distributional system would have raised output in the short term without any great need for higher fixed capital inputs.

The lack of incentives to participation in state and collective activity exacerbated the process by reducing the domestic supply of goods to the state-regulated distribution system. Low productivity in state industry thereby cut supplies of goods in the hands of the state for exchange with both its own employees and the agricultural cooperatives. An erosion of the role of the collective in agriculture because of the greater rewards to be had in "own-account" activity reduced the possibilities for procurement of food staples. The resulting pressure on the level of deliveries to state employees as rations further increased the demand for food on the free market and so the relative unattractiveness to cooperators of participation in collective labor. These arguments are sufficiently convincing to suggest strongly that the material incentives to full and effective participation in the agricultural cooperatives were weak throughout most of the period from the early 1960s to the mid-1970s.[27]

Statistical estimation of the parameters of the "outside" economy (kinh te ngoai) and its pervasive effects is extremely difficult. In the mid-1970s the income flows generated by such trade in staples seem to have been of the same order of magnitude as state-regulated trade, primarily because of the large price differential. The sheer size of the phenomenon suggests that almost everybody participated in it to some extent: cooperators active on the private plots or state employees "leaking" commodities onto the free market (small-scale embezzlement of state-distributed products, sale of secondhand goods, moonlighting on second jobs, petty-commodity production, and so forth). An important related issue was the use of administrative power to extract resources from the outside economy by the sale of licenses or by selective nonenforcement of the law (by turning a blind eye toward such activities in return for a consideration). The basic characteristic of such a system was the built-in structural competition for resources between the socialist and the outside sectors. This stemmed at root from the use of administrative measures to override immediate material incentives.

In the mid-1970s North Vietnam remained, by almost any criterion, at a very low level of economic development. This was despite the construction of the orthodox administrative apparatus of a "proletarian state" and considerable advances in such areas as education and welfare. While reliable per capita aggregate income data are unavailable, the following indicators provide some confirmation of this point:

1. Industrial employment in 1975 totaled 0.9 million—3.7 percent of the total population. Of these, perhaps 0.35 million (1.4 percent) were active in the modern plants largely run by the state, with the rest in artisanal and other small-scale cooperatives, as well as the residual "individual" sector.[28] Population growth far outstripped the employment-generating powers of the industrial sector (see table 1.1).

2. Crude unadjusted official data on food production show a gross annual output of only 5-5.5 million tonnes paddy equivalent in the period 1973-75, for a population of 24.5 million that was still largely dependent upon subsistence agriculture. This gives an apparent gross availability, before taking account of imports, of around 12 kg per capita per month on a crude milled-rice equivalent basis. Agricultural exports were minimal.[29]

Table 1.1

Reported Absorption of the Rising Labor Force, 1960–1975

Increase in population of working age	+3.45	100 percent
Increase in identified employment	+1.32	38 percent
Of which: State industry	+0.24	7 percent
Nonstate industry	+0.15	4 percent

Source: Based upon KTVH, tables 10, 12, 37, 68, 73, and 97.

3. Official data on the expenditure patterns of state employee households show that over 70 percent of outlays went on food in the mid-1970s.[30]

To the adverse relative price incentives created by the widespread outside economy should be added the problems of underemployment. Table 1.1 gives official data on the reported absorption of the rising labor force, 1960–1975. Of a reported increase of 3.5 million in the working-age population between 1960 and 1975, only 1.32 million (38 percent) went into identified employment.

This conflicts with evidence from official and other sources of labor shortages as a result of the heavy mobilizations to support the war effort. To the extent that the highly seasonal demand for labor in agriculture led to intermittent crises, this may have been true. But the mobilization of labor was also dependent upon the economic alternatives available to the individuals concerned, which early made own-account activity attractive.[31]

Production Conditions

Since I am neither a farmer nor an agronomist, I cannot really judge properly the conditions of production faced by North Vietnamese peasants. Two overwhelming impressions do, however, stand out. First, there is the great natural beauty of the landscape, especially its lighting. Second, there is the phenomenally high concentration of people. A commune of perhaps 5,000 people with around 300 hectares could be within around 2,000 meters of its neighbors. Yet within it life would have its own particular rhythm and character. By the mid-1970s the direct physical effects of the wartime bombing and straffing had been to some extent mitigated by the passage of time. The period of most sustained and violent attacks had been 1965–68, which had ended with U.S. President Lyndon Johnson's "unconditional bombing halt." The raids on Hanoi and Haiphong ordered in the build-up to the signature of the Paris Agreements and the infamous Christmas bombings of 1972 were, by comparison, far less destructive to the rural areas.

The basis for the extreme population density—around 1,200 per sq. km in the lower Red River provinces—remained intensive rice cultivation. Most fields were sown to rice. This was by far the dominant part of the major branch of rural production. There were three such branches: the cultivation branch (*nganh trong trot*), the livestock branch (*nganh chan nuoi*), and the subsidiary branch (*nganh nghe*). This classification system had a more than simply statistical significance, as should become clear later on.

Despite substantial (if unquantifiable) efforts, wet rice cultivation retained its historical dependence upon precipitation.[32] Monsoon-based water supplies remained unreliable: the average rainfall varied greatly, as did the timing of the rains. Typhoons could deposit around one-half to one-third of the yearly average of rain in twenty-four hours.[33] Extremes continued to occur, with flooding, droughts, and high winds adding to production difficulties. The sheer violence of the Red River made it difficult to improve water supplies. While attempts had been made to develop irrigation works at the supracommune level, much time still went into moving water by hand (with various sorts of scoop) over relatively short distances within the commune or lower levels such as the village or hamlet.

The weather determined the production cycle, subject to changes in water-supply conditions resulting from water-works and variation in the crops and seeds used. Two rice crops were grown, and a further nonrice staples crop, though with geographical variations, as much because of differences in social organization as from climate. The basic rice crop, which was almost universal, was known as the Xth month crop, or, more literally, "the harvest." It was usually reaped around December after transplanting in July-August, and it depended upon the monsoon rains of midsummer. Where possible a second rice crop was grown on the same fields. Derived historically from faster maturing seeds developed in the now defunct country of Champa, from whence its name probably came, this was conventionally described as the Vth month or spring rice. Harvested in May-June after transplanting in January, it was more dependent upon irrigation. The potential gap between the harvesting of the Xth month crop and the transplanting of the Vth month crop could be used for attempts to extend the area of nonrice staples by growing a winter crop. This frequently involved shortening the growing period for the Vth month crop.

There was some expansion of the area sown to new rice strains, which can be seen in the aggregate statistics showing the rising yield to the minor Vth month rice crop. This increased from an average of 1.74 tonnes per sown hectare in the period 1960–65 to 2.36 in 1973–75 (see appendix table II.1). The new rice seeds needed well-irrigated fields, and there were considerable problems getting them to adapt to the particular climatic conditions of the Red River Delta.[34]

Nonrice staples grown included maize, manioc, and potatoes (which presented considerable problems of disease)[35] and tended to be referred to together as the winter crop, for obvious reasons.

Supplies of fertilizer were of two kinds: limited amounts of modern chemicals and a variety of organic fertilizers among which azolla, grown on the surface of paddy-field water, was an important supplement to manure. Draught power was either human, water buffalo, or machine. Manure from draught animals supplemented that from pigs, which, with poultry, were the most prevalent types of livestock. Household gardens were a prime user of animal manure, but this varied with social organization. Pigs tended to be reared by households, using human waste, or in larger herds organized collectively. In the latter case foodstuffs for them might be grown specially, often using part of the winter crop, but sometimes rice was supplied.

The subsidiary branch included a considerable variety of nonagricultural activities. Both the style and manner of work varied between cooperatives. Certain lines, such as brick and tile firing, or mat weaving, often depended upon outside inputs. The quality of housing in the late 1950s had been very poor indeed, and in many communes tiled roofs and brick walls were a rarity. There was a correspondingly great demand for building materials, and many communes sought to have their own kilns. Artisanal activities had survived the French period relatively intact because of the low levels of peasant cash income and the lack of colonial interest in preservation of a mass market. Small metal goods had become of increasing importance as materials became easier to find during the First Five-Year Plan. Basketware, hats, sleeping mats, and other goods made from rushes and similar crops were also made.

The available official statistics on the balance between the three branches of production included the residual section of the peasant population (around 5 percent) who remained outside the cooperatives. Throughout the period 1960–1975 "subsidiary" activities contributed between 9.5 percent and 11.5 percent of gross agricultural output. Cultivation retained around 70 percent of reported total gross output throughout the same period.[36] These data are not likely to be very reliable, not least because of the use of state prices at a time when free market prices were both higher and far from stable. They do show, however, that diversification out of cultivation and into nonagricultural or livestock lines could not have been very large. The powerful effects of free market prices at the margin were extremely important for livestock rearing, above all pigs (see chapter 8).

The development of the "social infrastructure" had deep and pervasive effects. The centralized and active state apparatus played a major role. After the literacy drives of the late 1940s and 1950s the growth of the mass media—primarily radio, newspapers, and journals—greatly changed the farmer's environment. The literacy of the general population had greatly enhanced the centralized state's ability to convey information, while the development of the mass media made possible advance warning of typhoons, dissemination of techniques for controlling disease, and so forth. The extensive educational system facilitated health improvements, for example, through measures of personal hygiene, so strengthening the system of basic medical care.

But apart from the social infrastructure and the direct effects of the war it is hard to believe that production conditions in North Vietnam were so very unusual by comparison with other extremely poor rice-growing areas. Population density was very high, industrial inputs were very low. Use of new seed strains was not

Table 1.2

Key DRV Agricultural Indicators

Annual averages:	1960–65	1973–75
Staples cultivation		
Rice: area sown (million hectares)	2.38	2.20
Yield (tonnes/sown hectare)	1.82	2.23
Output, million tonnes paddy	4.34	4.91
Nonrice staples: area sown (million hectares)	0.51	0.43
Yield (tonnes/sown hectare)[a]	1.67	1.72
Output, million tonnes[a]	0.85	0.74
Total staples output, million tonnes[a]	5.19	5.65
Total population (million)	17.4	23.9

Source: See appendix 2.

[a]Crude milled rice equivalent.

yet extensive, and was in any case limited by lack of fertilizer. It should be remembered, however, that the area started from an extremely low base, and produced inputs and livestock use had both been cut back further during the colonial period.[37] But—as shall be seen—increased inputs at the aggregate level were recorded, so that the outcome in terms of slow growth in yields and output was rather surprising.

Agricultural Output

Analysis of agricultural development in the DRV (table 1.2) is complicated by the effects of the two periods of bombing and straffing (1965–68 and 1972). Year-to-year variation resulting from the weather can be partly offset by appropriate periodization, but conclusions drawn cannot be decisive.[38] Detailed statistics and a simple periodization can be found in appendix 2.

Rice. Official data on paddy output showed a 10–15 percent increase between the two periods 1960–65 and 1973–75, although at times the data may have overestimated total output by the same proportion (see below). The gains report- ed arose from higher yields on a reduced area. Even allowing for the difficulties involved in bringing war-damaged land back into cultivation, the Xth month area in 1975 was still over 10 percent below the peak year of 1964. A possible additional explanation for the apparent decline in cultivated area is encroachment upon cooperative land for own-account cultivation by cooperators.[39] Rather than report an excessive rise in the private area, local officials explained away a fall in the collectively farmed area by reporting a reduction in the cultivated area. An additional cause may have been the desire to maximize reported yields in order to obtain plan-fulfillment bonuses. This would encourage a reduction in the reported area for a given level of reported output.

Secondary staples. After a peak in the production of nonstaples during 1967–68 these crops were produced in lower volumes on lower areas. The fall amounted to approximately one-third of the increase in paddy output.

Nonstaples. The data divide this group into four—vegetables, yearly industrial crops, perennial industrial crops, and perennial fruits. Only nonbean vegetables developed significantly between 1965 and 1975. North Vietnam's comparative advantage in the socialist bloc as a producer of tropical products necessarily remained largely underutilized, although during the First Five-Year Plan tea, coffee, and rubber output all grew rapidly as the state farm sector expanded.

Livestock. Both pig and cattle numbers grew rapidly, and this was one of the few obvious successes of Vietnamese agriculture. Almost all of the advance, however, occurred in the household sector. This was quite contrary to party policy, which emphasized the role of the cooperative and the need to set up centralized collective herds.

Agricultural Inputs

The general picture of stagnation sits rather oddly with the data showing input increases. In some areas—such as pumps (see appendix table II.5)—these were rather rapid, especially after 1965. It is also surprising that cattle and water buffalo numbers appear to have been higher during 1966–68 than during the previous six years (appendix table II.3). Yield stagnation is to be explained, as this study tries to show, by resource utilization problems. Unfortunately there is very little information regarding aggregate resource use.

Area. Attempts to estimate the potential cultivable area are unrealistic at present. The delta landscape changed substantially after cooperativization as amalgamated fields replaced the traditional field pattern. This process marginally increased the available area by removing uncultivated borders.

Water control. This arguably remained the crucial input given agriculture's relatively low level of development. Data (appendix table II.4) on the availability of pumps and water-works constructed by the state show that the numbers and areas involved were not unappreciable. While the damage done by bombing cannot be ignored, at a minimum some 150,000 hectares were being served by pumping stations in the mid-1970s—well over 10 percent of the paddy-field area.

Mechanization and electrification. This apparently progressed to a surprising extent, but primarily in such areas as threshing: ploughing remained almost entirely unmechanized.

Fertilizer. The growth in animal numbers implied some increase in the volume of manure available, but there are no aggregate data on fertilizer use. Because the pigs were largely kept in private hands, most was probably used on the private plots. Use of azolla had spread because the DRV had continued dissemination of methods used to grow it, for a monopoly on knowledge of its use had previously been held by certain districts. Chemical fertilizer was in great demand, and production rose nearly eightfold between 1960 and 1975, by which year crude tonnage was 0.4 million tonnes.[40]

Labor. Although labor availability had risen with the increase in population there was no DRV measure of active employment in agricultural activity. Cooperative membership was no guide to collective labor inputs (see chapter 7). The reported average number of workers per cooperator family fell between 1960 and 1965 and then declined sharply in the late 1960s. It started to recover in 1970 and rose slowly in following years, but not back to the levels of the early 1960s.[41]

Judging the precise levels of agricultural inputs at the aggregate level is therefore almost impossible. The information available shows, however, that there were some significant increases in gross resource availability, primarily labor, fertilizer, and seed varieties. This was not sufficient to lead to a substantial rise in output, and a decline in aggregate per capita staples output had to be offset by a shift toward structural dependence upon food imports. This shows up in the overall food balance of the area.

The Food Balance

Stagnation in aggregate food output resulted in growing difficulties in balancing food supplies at the level of the aggregate economy. Population growth continued to be rapid, averaging 2.8 percent annually over the period 1960–1975.[42] This gave an added twist to the cumulative macroeconomic interactions involved in the wake of food shortages, most especially the growth of the free-market, "outside" economy and the accompanying continual shortages within the admininistered "socialist" sectors.

According to the official data, domestic food production did not grow sufficiently to cope with the increased number of mouths to be fed. On the basis of official statistics and taking the widely accepted basic rationing level of 13 kg of crude milled rice equivalent per capita per month, the situation had apparently reached a crisis point by 1968, by which time reported food supplies were biologically insufficient. Substantial imports from the Soviet bloc began in that year and continued throughout the 1970s. The position is made less certain, however, by the absence of reliable data on Chinese food aid.

There is also considerable doubt about the year-to-year accuracy of official output data. In some years the official figure for domestic staples production implies a rather substantial surplus above subsistence retained by the cooperatives while imports continued.

Another difficulty arises from the fact that although there is evidence that state employees increased the volume of their purchases from the free market, the estimated aggregate surplus above subsistence in agriculture did not rise. The estimated volume of sales to the free market by cooperators showed some growth,

Table 1.3

**Estimates of the DRV Food Balance
(million tonnes of crude milled rice equivalent)**

Year	Production	Estimate of net imports	Per capita availability[a]
1960	3.0	Nil[b]	16
1965	3.6	Nil[b]	16
1974–75 average	3.8	0.5	15

Sources: Net imports—see text and appendix table I.2. Production and population data—see appendix 1.

[a]Kg per capita per month.
[b]Estimated at less than 0.1 million tonnes; this is an extremely approximate calculation in the absence of reliable data.

but there was an apparent standstill in real cooperator incomes from the household economy. It seems likely that sales onto the free market of surplus rations and products "leaked" from the state distribution system were a further complication. In addition, reports of private plot production are based upon simple sample surveys and are almost certainly gross underestimates. Thus the available indicators on the likely balance of food supply and demand at the level of the entire economy are somewhat contradictory and far from unanimous in the picture they reveal.

There are good reasons for not trusting official data. Reports on free-market participation were highly contentious and politically loaded. Output data have both upward and downward biases: evidence from the micro level shows the existence of incentives to underreport output in order to reduce the procurement liability, while plan overfulfillment bonuses would act as incentives to overreporting.

Two fundamental problems remain for any analysis: first, the level and timing of Chinese food aid; second, the sources and channels that fed the free market in staples. But almost nothing is known about the two other crucial areas—the policy toward food stockpiles, and interregional (especially interprovincial) differences. Table 1.3 presents some estimates of the DRV food balance.

Conclusions

Reported food output growth in the DRV over the period 1960–1975 was inadequate compared with the rise in population. High levels of imports were needed to meet the gap between domestic production and needs that opened up in the late 1960s. Additional inputs into agriculture were far from negligible, but these were offset by the major U.S. bombings of 1965–68 and 1972. There was very little diversification away from either rice or staples cultivation. The macro economy

Table 1.4

**Apparent Surplus above Subsistence in DRV Agriculture
(million tonnes crude milled rice equivalent)**

Year	A: Total staples production	B: Estimated subsistence needs of cooperator pop.	C: State procurement	Apparent surplus = A-B-C
1960	3.0	1.8	0.6	0.6
1965	3.6	2.1	0.7	0.8
1974–75 (average)	3.8	2.5	0.6	0.7

Sources: KTVH, tables 70, 71, and 93, and see appendix I.

became characterized by inflation and the growth of tensions between a high-price "outside" sector oriented toward the free market and the administratively managed and import-dependent socialist sectors. Much of the aggregate data is inconsistent and not very reliable.

This describes the basic parameters within which the agricultural cooperatives operated in the mid- and early 1970s. Policy makers saw stagnation and, it is clear from the micro sources, great waste of resources. They also saw systemic competition from the outside economy for economic resources and a corresponding tension between cooperators and the collectives. At a minimum, the cooperatives were seen as suppliers of staples to the state under the various forms of procurement. It is therefore vital to realize that procurement was, in fact, a high proportion of reported output through the period. In this area the collectivized system was in practice not a failure.

Procurement

The procurement of agricultural produce was, and was intended to be, a major element of cooperative-state relations. Substantial volumes were obtained, and here the data may be more reliable than elsewhere.

Table 1.4 shows the apparent surplus above subsistence in agriculture after deliveries to the state. Gauging the pressure of procurement upon cooperators' living standards is clearly of great importance, but it is extremely difficult. To estimate it accurately would require knowing the average level of collective distribution of staples (i.e., net of taxes and sales to the state as well as seeds, etc.), and this is not available. The evidence, such as it is, suggests a steady decline. Whereas before the crisis of the late 1960s this was largely because of pressure from procurement, subsequently there was probably a general shift toward private plot production at the expense of the collective; this very possibly led to an improvement in cooperator real consumption. Comments from Vietnamese economic writings after 1979 have hinted at average levels of distribution near the 5–6 kg per capita per month level.[43] This is well below subsistence and would suggest that a major part of the immediate attraction of own-account activity was to meet biological requirements.

According to the official data there was probably a substantial surplus above subsistence needs in the cooperatives after deliveries to the state. Note that the output data in principle include those from the private plots which were (in principle if not in practice) limited to 5 percent of the cultivated area, as well as the state farms. But the same statistics also imply aggregate per capita availabilities well in excess of subsistence in the mid 1970s, despite continuing high levels of imports. And although official data show an increasing volume of free-market purchases by state employees, this is most unlikely to have come entirely from suprasubsistence levels of distribution to members of agricultural cooperatives. The micro-level evidence suggests that such levels existed in only a small minority of cooperatives.

Some further idea of the pressure of procurement upon cooperators can be obtained from the production mix, which implies that pressure rose up to 1967–68 and then declined. This conclusion relies upon the phenomenon of "producing-down"—when inadequate subsistence is to be had from existing output, recourse is made to lower-grade staples that provide a higher nutritional yield per hectare, such as maize, manioc, and potatoes. Pressure upon biological requirements leads to a lowering of the quality of the diet. Thus while official policy encouraged the production of these subsidiary crops, the rise in their area and subsequent decline after 1967–68 suggests an improvement in the quality of the diet during the later period consistent with a reduced pressure on real incomes, perhaps due to an easing of procurement. It also seems to indicate that the apparent crisis of 1968, resolved by a structural shift to dependency upon food imports, also saw the passing of a peak of pressure upon cooperators from the state. The evidence from the net barter intersectoral terms of trade also suggests an improvement of the position of agricultural producers in the early 1970s, with rapid reported rises in state purchasing prices.[44] But a fundamental problem here (apart from the weakness of the data) remains the lack of detailed information on goods supplied to agriculture, which precludes the drawing of any firm conclusions concerning cooperative-state relations.

The inferences drawn from the above are on the whole consistent with the argument that the growth of the high-price outside economy and the associated phenomenon of shortages within the state-controlled distribution system tended to reduce the relative incentives to participation in collective activity. This also benefited those with access to marketable resources at the expense of others, such as state employees who could not generate income flows outside. Whether for subsistence or for sale on the free market, own-account production on the private plots was an attractive alternative to collective labor for most cooperators during the mid-1970s. Nominalization of their cooperatives was one way of realizing that alternative, and such broad-based opposition to official policy was a fundamental part of the North Vietnamese Agrarian Question.

Some Preliminary Issues

The effectiveness of cooperative institutions as vehicles for economic development depends greatly on both the problems they are intended to solve and the environment within which they have to operate. Vietnamese historical experience suggests that, as local institutions vested with substantial formal property rights, cooperatives would be expected to provide minimum levels of income support to offset both the highly risky environment and the substantial variation in private economic resources of households, especially with regard to labor power. In addition, they would be seen as the legitimate objects of social competition to attain positions of authority in such local corporate bodies.

Superimposed upon such static conceptions, however, was the party's commitment to social advance and national economic development through the Vietnamese Socialist Revolution. As will be seen, this posed additional questions about the relationship of the cooperative to the rest of the DRV economy, and especially the effects of any individual's participation in it upon the revolution. This naturally shifts discussion away from simple developmental issues, for, especially in wartime, issues of nationalism, of national advance and development, and of the position of the individual in the wider sphere are of considerable importance. Such arguments had a bearing upon the legitimacy of the party and state in peasants' eyes. Here it is important to note that, just as paying taxes would not necessarily imply an acceptance of the right of the authorities to levy them, so the frequent opposition of local interests to official policies does not necessarily suggest that peasants saw their proponents as illegitimate occupiers of positions of authority.

These questions are important. The impact of ideas of national advance and socialist construction upon peasants may lead them to disregard—if only partially—their immediate material interests. Despite adverse prices, comparatively large levels of procurement did, in fact, occur. The fact that the collectivized system accompanied high supplies to the state of men and food throws revealing light upon the criticisms policy makers made of it in the late 1960s and early 1970s (see chapter 4). From these criticisms came the NMS and the commitment to it reaffirmed in 1976 at the same time as the de jure reunification of North and South. Thus it starts to become clear that developmental issues of growth and surplus procurement are not sufficient to explain the value attached to the NMS and the collectivized system. In the early 1970s North Vietnamese collectivized agriculture was meeting basic targets of procurement and troop mobilization, but the collective part of it was, as will be seen, economically inefficient. The Agrarian Question was to be found in issues relating to the role of the plan, the value of Marxist-Leninist theory in the unified management of the country, and, most importantly, the day-to-day flaunting by the mass of the peasantry of the party's norms for cooperative management.

Part I

The Higher-level Agricultural Producer Cooperative

2

The Producer Cooperative in Rural Society

With nearly 90 percent of the population living in rural areas during the mid-1970s North Vietnam remained a society largely dominated by people from peasant backgrounds. The various positions in the communes and cooperatives were the base upon which the party and state structures rested. Their occupants, the rural cadres, were responsible for such essential tasks as tax deliveries and troop mobilization. The strength of their position and their close relationship with other peasants helps to explain the stability of the society during the war against the United States. The state farms, which employed only 2 percent of the agricultural labor force in 1975,[1] were peripheral to this system and were mostly concentrated in the upland plantation areas. The most important problems arising from rural society came from the collectivized communes of the delta rice-growing regions.

The administrative system used in the 1970s was not a radical departure from previous experience. A hierarchy extended down to the local level from the central organs in the capital. The basic unit in the countryside continued to be the commune, where there was both an administrative People's Committee and a Party Committee grouping the local cells. But the existence of collective economic units (primarily the agricultural cooperatives) beside these structures inevitably led to confusion. In addition, all these formal official hierarchies had to deal with more basic underlying power structures. These were often identified with groupings below the commune level. It should be noted that the local administrative-geographical system had three levels, here translated for convenience as the commune (*xa*), the village (*thon*), and the hamlet (*xom*). All of these were basic population units that had their own distinct interests, and they will enter frequently into the discussion. It should be stressed that, contrary to Chinese practice (1958–1982), the commune was an administrative level quite distinct from the cooperative.[2]

A village was usually a continuous area of habitation. It was typically divided into neighborhoods (hamlets) which were often subregions based upon side-streets off a main street. Hamlets were less well defined than villages. Villages were formally grouped into communes, and they did not usually have direct relations with higher levels. A typical commune of 5,000 people might contain two or three villages and a dozen or more hamlets. Because of changes in administrative boundaries communes did not always correspond to dynastic units and occasionally consisted of villages from a number of the older communes.[3]

The Vietnamese words for cooperative and cooperator are literally translated as "united-task-commune" and "commune-member."[4] This provided linguistic continuity. In addition, the cooperative took over some of the dynastic commune's traditional functions. Tax collection, labor mobilization, and social welfare activities were the most notable. Cooperatives tended to be based upon the population groupings associated with preexisting administrative units. Initially these were simply hamlets, but then, as the cooperatives grew in size, villages or communes. The growth and population catchment area of the production brigades (the cooperative's most important subunit) followed a similar course. Thus the process of amalgamation associated with attempts at managerial reform could not avoid involvement with patterns of local interest. This had considerable influence upon the outcome of attempts to implement prescribed cooperative management methods.[5]

The pressure of population in the Northern deltas and a tendency to build upon higher land to avoid floods led to a high housing density within villages. Average distances between communes were not great, which meant that markets were within easy reach. The level of administration superior to the commune was the district (*huyen*), and above that the province (*tinh*). It would normally be possible to walk to the district capital in the morning and return for the mid-day meal.

Thus, taking the administratively recognized family (*ho*) as base, there were seven levels: center, province, district, commune, village, hamlet, and family.

The formal relationships of cooperatives with higher levels were understood to be as follows.[6] The main regulatory agency was the Agricultural Office at the district level, supplemented by offices responsible for such areas as planning, cadre organization, and basic construction. It was presumed that these also included an office exercising financial regulation on behalf of the National Bank, which attempted to limit the cash held by the cooperative.[7] The district also possessed a local office of the state purchasing organization, which was based upon the province. In practice the primary sources used here had little to say about such relationships, which suggests that they were not very important,[8] despite the emphasis placed upon them by the central authorities (see chapter 4).

It was the party's customary practice to have a parallel system of party committees monitoring the state apparatus. This provided a channel for influence both out of and into the cooperatives that could be used to bypass the state apparatus if and when necessary. (See the discussion in this chapter of the "democratic nature" of cooperatives and the question of the "agency of reform" in My Tho cooperative in chapter 6.)

Rural cadres held positions in a variety of mass and party/state organizations: the party cells, the party committee(s), the people's committee(s), the military organizations, and the various associations of the umbrella Vietnam Fatherland Front—women, youth, and so forth. In addition, they occupied key posts in the production brigades and other organs of the cooperatives. Although differences between these formal positions of authority corresponded to different functions and sources of power, the grid-like structure of overlapping organizations dominated by the party was intended to reinforce internal cohesion.[9] As a whole, the group possessed very strong sources of power and authority: control over all direct economic relations with the state; control over collective production and output; regulatory control over the household sector; and the police.

In any commune this group probably numbered between twenty and fifty. In addition to the commune's committees, the cooperatives typically had a dozen brigades, each with a head and a deputy, and perhaps ten to twenty people in the various committees and central managerial organs.[10] The internal unity of this group was a major issue. Clearly it had some collective self-interest; in addition, the party leadership placed great emphasis upon its maintaining internal cohesion. Despite this, however, many basic phenomena, especially those accompanying managerial reform, can only be explained in terms of a fundamental split between brigade leaders and the occupants of posts in the higher management organs of the cooperative. The origins of these divisions are discussed below.

A widespread contention among certain North Vietnamese was that the simple reason why cooperatives "did not work well" was that cooperatives were political units as well as vehicles for economic development. This assertion will be reconsidered in subsequent chapters. From this point of view, the key point about the cooperatives was that while they did not provide the sole basis for leadership in the communes, they were nevertheless used as such by rural cadres, for whom they were vitally important. This sought to explain simply a whole range of problems typically labeled "conservatism"—for example, the nonadoption of apparently advantageous and relatively costless technical advances. The explanation for the nonimplementation of official policies given to me by an agricultural economist followed this line of argument: "Cooperatives are at the same time economic and political units. Because of this the leaders of the cooperatives are, generally speaking, party members and have shortcomings with regard to economics. Because of this the function of the 'office of help' [a team sent in from outside to assist with managerial reform] is to help the Management Committee.'' Or, ''the *conservatism* of the old party members who are cooperative Chairmen. Usually, in a production brigade, the leader or deputy are party members.''[11]

This view suggests that one should not try to explain economic phenomena in isolation from these factors. An immediate rejoinder to such arguments is the question of why such people kept their positions. Such ''incompetence,'' if matched by a ''liberal'' attitude toward ordinary cooperators (e.g., toward the use of cooperative assets), might still retain local support.

The continual overlapping of functions in a wider leadership role exercised by dominant individuals is of great importance. Such broad concepts are vitally necessary because of the complexity and wide range of social interactions. When different sources of socioeconomic power coexist, as they had to in a collectivized system like that in North Vietnam, underlying power structures can surface in the way people use organizations for quite different purposes from those formally intended. This is well illustrated by an imaginary example.

I asked a Vietnamese expert who was familiar with practical problems how the following social problem would typically be resolved: if two old women were in dispute over a banana plant situated on the border between their yards (a dispute that was rapidly becoming acrimonious and ruining everyone's sleep), who would be responsible for providing a solution? According to the reply, rather than some type of administrative organization, the production brigade would make the first attempt at reconciliation—that is, the unit for which members of the two old women's households worked when working for the cooperative. Only if this was unsuccessful would the communal administration's formal procedure be used—not, it should be noted, the cooperative's.

This example illustrated the importance of subunits of the cooperative in an area of economic life that was, strictly speaking, outside the collective sphere. Presumably, though, the reason the brigade leadership would be called in was that they were the most suitable people to deal with a problem that would clearly require authority, if not considerable tact. Such phenomena would have had consequences for the two households' overall relations with the brigade leadership. Because of this there would also be some impact upon the state of their relations with the socialist production unit stressed by central policy-makers—the cooperative. In principle but often not in practice, the brigade was meant to be closely controlled by and dependent upon the Management Committee of the cooperative. In this example the production brigade—in principle a subordinate part of the cooperative (the collective economic unit)—was providing a focus for the social authority needed to regulate ''private'' economic relations. Thus the particular ''private'' interests of the two households (the price of bananas, the beauty of the granddaughter) could affect in some way the balance of relations between brigade and cooperative.

Formal Structure of Cooperatives

Discussion now centers upon the economic role of a cooperative in rural society. Because conflicts often centered upon the precise nature of this role it is essential to examine in detail the formal structure laid down in the official Statute. This specified the formal property rights and the duties and responsibility of agents within the system. In practice collective activity could well be organized by the brigade, not the cooperative, while cooperator income could depend more upon the results of household than of collective production. Later chapters develop a more realistic analytical approach which is used to explain the enormous variations between cooperatives, the vast discrepancies between principle and practice, and the actual significance of such terms as "collective property" or "unified economic management by the cooperative."[12]

Statutory Nature of a Cooperative

The authorities laid down the prescribed form of the higher-level agricultural producer cooperative (*hop tac xa san xuat nong nghiep bac cao*; referred to hereafter as cooperative[13]) in a detailed "Draft Statute." Different versions of this are available, which change in only minor respects over time. I have three copies of such statutes for "higher" cooperatives, which is the type relevant here. References in the text below to "the Statute" are to the version translated in Fforde (1984). Others are to the later 1974 Draft Translation (by articles).[14]

The basic attitude of all three was identical: they defined the constituent elements of the social organization of the cooperative and then allocated duties and obligations to them. Social relations were seen as harmonious, without emphasis upon potential conflicts of interest.[15] This is clear from the definition of a cooperative:

> The higher-level agricultural producer cooperative is a socialist collective economic organization freely set up by the working peasantry under the leadership of the party and state, which aims to increase agriculture and other production, overthrow exploitation, improve living conditions, and contribute to leading the North toward Socialism while consolidating the base for the struggle to unify the country. (Clause 1)

The last phrase required cooperatives to function in extraeconomic areas. These included the war effort, and they were given responsibility for mobilization.[16]

The cooperative's political and social functions were in principle extensive. This can be seen from a later section on the cooperative's "political-ideological" and "cultural-social" duties. Here, under the leadership of the Party Committee (*chi bo**)[17] of the commune, the cooperative was specifically required to propagandize the position of the party, while teaching the "collective spirit" and the "spirit of enthusiastic work" and encouraging cooperators' democratic participation in management decisions. In addition, under the heading of "cultural-social work," the cooperative was to organize educational activities and ensure that work was available for all. It was to give assistance to the families of wounded soldiers or others in difficulties. The cooperative was also to pay attention to improving the general level of medical services, the needs of pregnant women, and the provision of creches.[18]

These two clauses reveal the extent and variety of a cooperative's functions. To implement them, the authorities used the Statute to help define the internal organization of the cooperative in terms of the duties and responsibilities of its constituent elements. The Statute recognized five different categories of agents: the cooperative itself; its General Assembly; its two committees (Management and Supervisory); its constituent subunits, brigades (doi*) and teams (to),[19] and the cooperators. These will be considered in turn.

Cooperators

Subject to the approval of the General Assembly, anyone who was not from an "exploitative class" or disqualified by legal process had the right to join the cooperative. Place of origin was not relevant. Specific clauses dealt with the right to work of the under-age, the families of ex-landlords, and so forth and also ensured that "weak" families were admitted.[20]

Cooperators (xa vien) had the right to participate in collective work and in the management of the cooperative (including standing for its offices). They could also "expand the minor household economy (kinh te phu gia dinh*) in conditions that did not interfere with the cooperative's management of labor" (clause 6). The 1974 Draft Translation specifically states that "cooperators cannot expand the 5 percent land area for side occupations" (article 9). Their corresponding duties were to follow regulations and the directives of the management and, while defending the cooperative, actively to encourage others to join (clause 6). In practice these formal rights had two important implications. First, the "right-to-work" clause gave considerable impetus to labor sharing. It thereby reinforced tendencies toward reestablishment of traditional collective functions of risk insurance. Second, it is important that cooperators had a quite explicit right to engage in own-account activity in the "minor household economy."[21] Major conditions for the latter were access to land and markets, and these too were made clear in both relevant versions of the Statute:

> The land of cooperators must become the common property of the cooperative,[22] but in order to take care of the private needs of the cooperators it is neccessary to set aside for each household an amount of land equal to not more than 5 percent of the commune's per capita average for each member of the household. (Clause 9)

and

The products from the minor household economy are freely disposable. (Article 19)

Some allowance was also made for orchards, ponds, and so on, which could in some cases remain owned and managed by cooperators who could retain the profits from them.[23] These resources were often of great value to cooperators because they could generate cash incomes and protein sources.

The right of any cooperator to leave the cooperative was slightly different in these two versions of the Statute. In practice this right was almost incapable of realization.

Cooperators have the right to leave the cooperative, but must give advance warning and wait until the end of the harvest. (Clause 7)

Application should be considered by the Management Committee, and the Assembly will give final approval.[24] (Article 7)

The family was a key basic unit of population control in North Vietnam. The Family Passbook (ho khau) was an important document. The Statute did not specifically mention the family, however, as a level intermediate between the cooperator and the brigades or teams. The household (gia dinh) was, however, recognized as an area of activity, and it received the "5 percent land" (dat nam phan tram). The children of cooperators, apart from having the right to work in the cooperative, also did not have to buy shares when they came of age (sixteen years). This gave further implicit recognition to the importance of basic kin groupings.[25]

Brigade and Teams

A section on "the organization and discipline of labor" dealt with the approved managerial subunits of the cooperative, of which the brigade was by far the most important. Well-established usage in the mid–70s tended to accept that the brigade and the team were, in principle, subunits immediately below the cooperative level. The latter were directly managed, while the former were controlled by a contractual system, for example the "III-point contract with penalties and bonuses" (hop dong ba chieu voi thuong phat*) signed between brigade and cooperative. Groups (nhom*) were more ad hoc units below the brigade/team level; teams, however, were sometimes constituent elements of brigades. The Statute defined these subunits as follows:[26]

Apart from those small cooperatives with less than around fifteen families, larger cooperatives should, according to their own conditions, divide up their cooperators into fixed brigades or teams and allocate to them land, livestock, and implements to use in accordance with the general production plan of the cooperative. . . . When necessary, they can be further divided up into temporary work groups.

The brigades or teams elect a leader (*doi truong**) and a deputy (*doi pho**) to take responsibility for the work of the unit. The leader must distribute work appropriate to the abilities of the workers in his unit, supervise and check up on the results of each worker, and then note down their points. After setting up these units, the cooperative needs to realize work-contracts with them, from small contracts (by "stage" or harvest) to large (yearly), and from contracts in points to those governing production levels and expenses. It is also necessary to fix the system of penalties and bonuses. (Clause 24)

It is important to note that by avoiding direct contractual relations between cooperative and cooperators the Statute forced the brigades to act as intermediaries between the cooperators and the central organs of the cooperative. These were the cooperative's key Management Committee and its Supervisory Committee, which played a minor regulatory role.

Committees

The Management Committee (*ban chu nhiem**) and Supervisory Committee (*ban kiem tra**) had extensive duties and powers. The former, headed by the manager (*chu nhiem**), was to regulate the cooperative's economic activities, execute the resolutions of the cooperative's Assembly (see below), and represent the cooperative in its relations with outside organizations (clause 36). The last inevitably recalls the importance placed in the dynastic commune on its role as intermediary between peasants and higher levels. In contradiction to the definition of brigades given above, both the Statute and the 1974 Draft Translation stated that the Management Committee could assign its members to act as brigade leaders. This limited the brigade members' formal rights to elect their own leaders. The Supervisory Committee had an audit function. It was responsible for overseeing the Management Committee and calling extraordinary meetings of the General Assembly. No overlap of membership was allowed between the two committees (clause 37). The Supervisory Committee's sole right to call extraordinary General Assembly meetings allowed for more effective control by leading local party cadres, for the secretary of the Party Committee was usually chairman of the Supervisory Committee.[27] Both these committees were formally subject to the cooperative's highest organ, the General Assembly of Cooperators (*dai hoi xa vien**). Arrangements were made, however, to limit severely the Assembly's formal powers.

The Cooperative and Its General Assembly

Because it was the basic instrument for implementation of official policy in the rural areas, the cooperative's formal aims were of great importance. A general statement was given as follows:

> on the basis of collective labor and technical progress ceaselessly to increase productivity and income, improve the living conditions of cooperators, accumulate and ceaselessly develop the superiority of cooperative to individual production; the key problem is raising productivity and incomes. In order to realize this, the cooperative must have a detailed plan, aiming to use rationally the labor-power of the cooperators, strengthen the development of production, and increase the volume of work done, raising the incomes of the cooperative and the cooperators (this must entail diversification of production). (Clause 22)

This set of policy aims has clear parallels with those common in the CMEA and noted in chapter 1. Because the Statute's authors held to their underlying assumption that conflicts of interest were absent, many important issues were ignored. For instance, there was no stated method for resolving such issues as the potential trade-off between work-sharing and improved labor productivity. If certain individuals had a higher social productivity than others, and therefore output was higher if they were allocated the limited amount of work available, then how were others to be compensated? Again, if a rich brigade was to give up some of its land as part of a "rationalization" of the cooperative's landholdings, then how should it be compensated? These were important social questions that the Statute left unanswered.

In principle the cooperators made their wishes felt through the operation of the cooperative's General Assembly:

> The highest organ of the cooperative is the General Assembly of cooperators. It has the following powers and duties:
>
> a. Examine and accept the work reports of the Management and Supervisory Committees.
>
> b. Examine and accept the production plan, financial estimates, work norms, and points-standards, and sign important contracts with the outside world.
>
> c. Examine and determine the price of livestock, implements, and other items when they are made into collective property, and value of shares, and review the way in which the product is distributed each year.
>
> d. Pass and correct the cooperative's Statute.
>
> e. Appoint and dismiss the members of the two committees.
>
> f. Admit new and provisional members to full membership.
>
> g. Examine and determine the various bonuses and heavy penalties for cooperators, and the expulsion of cooperators.
>
> h. Examine and determine other important work of the cooperative. (Clause 35)

Although regulations existed governing quora and the frequency of meetings of the Assembly, the lack of formal provision for the calling of extraordinary meetings by cooperators severely weakened the democratic nature of the system. If the cooperative's management organs were dominated by an unpopular but intimidating clique, corrective action was clearly dependent upon informal pressure from inside the cooperative or by some outside group. In practice, I was told, such problems (e.g., embezzlement by cadres) would be dealt with by the party.[28] The clique would eventually be forced to call an Assembly meeting to remove them. If the party secretary was, following custom, chairman of the Supervisory Committee, then this presented clear dangers—and opportunities. The custom could possibly have had the benefit of allowing younger men to be managers.

Supervision and control by the party carried over into more ordinary relations with the outside world:

> The party cells in the commune were always present at the level of the cooperative, united at commune level in the Party Committee. The commune in those situations where cooperatives were quite small was the administrative and political level—if taxes were allocated by the district to the commune, then the president of the commune would ensure that they were paid . . . although responsibility would rest with the cooperatives. Taxes were delivered by the cooperatives directly to the district, not via the commune. (Hanoi I)

This shows once again how misleading it is to separate economic from noneconomic functions. As part of the "wider leadership role" mentioned above, important individuals clearly acted so as to use their general authority in a number of different areas at once. This did not fit neatly into the division of responsibilities envisaged by the formal organizational structure of the collectivized system. For party cadres this could appear to have immediate benefits, since the economic functions of the cooperative could thus depend upon political organizations—in this case probably the party's commune-level committee. As will be seen, the attempt to separate responsibilities by making the cooperative liable for tax payments clearly did not work in practice.

The cooperative's responsibility for ensuring a planned development of production has already been mentioned. In addition it was meant to use better practice techniques, accumulate directly through the use of its own resources, and invest in areas such as livestock and tools. But the most basic requirement of development was the cooperative's plan:

> Each year, toward the end of autumn, the cooperative must construct a production plan for the coming year on the basis of past experience and future duties (*nhiem vu*). The contents of the plan must include: the general level of production and the levels for each product, technical means, a plan for hiring manpower, livestock and implements, supplies of means of production, a plan for the

consumption and transformation of products, basic construction, etc. When setting up the production plan for the cooperative, the Management Committee must allow the cooperators to participate in a democratic way, and avoid coercion and authoritarianism. (Clause 23)

Under the New Management System the plan should also have had distributional elements. These included the estimated value of the workpoint and procurement levels. In the Statute, however, attention was directed to the establishment of funds (*quy**) to finance accumulation and social expenses. The prescribed method of output distribution after the harvest was as follows. After allocating taxes and making various sales to the state, the cooperative had to meet its own administrative expenses. These were not meant to exceed 1 percent of gross income. The next calls on income were the nonlaboring income recipients and the funds. Finally, "the rest [could be divided] among the total number of labor-days worked by cooperators during the year." Note that this was usually given out all at once after the harvest, and not monthly.[29] Practice varied.

The funds were of considerable potential importance:

The accumulation fund (*quy tich luy**) is used to buy draught animals and tools, to pay for basic construction such as the building of warehouses, waterworks, expand the area, clear forest, dig ponds for fish, etc. . . . and can also be used for the finance of production expenses when absolutely necessary. At the beginning this should be allocated 5 percent of income [i.e., production less expenses]; when production has developed this can be raised to 12 percent.

The social fund (*quy cong ich**) is used for cultural and social work in the cooperative. At the beginning contributions should not exceed 1 percent of income, but later they can be raised to 2–3 percent.[30] (Clause 32)

There was extensive provision for the allocation of resources to "nonworking members." These were of two basic types: the cadres and those in receipt of welfare payments.

Cadres of the cooperative such as the manager and the accountant, because they are busy with the cooperative's work, cannot do very much [manual] work. Therefore the Assembly must, yearly and based upon the work of each man, fix for them the minimum number of days they must [register for] work and allocate them a definite number of workpoints so that the number of days they work [i.e., receive workpoints for] is comparable to that of the average cooperator. (Clause 27)

Such cadres should also have been subject to a system of penalties and bonuses related to the degree of plan-realization.[31] In addition,

Families [of war-dead, wounded soldiers, sick soldiers, *bo doi*[32]] who are short of labor-power need to have the cooperative arrange work properly for them. They should receive extra workpoints in addition to those they actually work for. The resulting total should be equal to that of an average family. *This clause can also be used for the families of cadres of the commune such as the secretary of the Party Committee, or the president of the commune.* (Clause 28; emphasis added)

The latter point applied because "they [the cadres] are busy with common work."

The great mass of collective income was meant to be distributed through the system of remuneration for work done. Apart from stipulating that meeting it was a duty, the Statute said little about either the level of procurement or how it should be determined. The principles governing cooperative-state relations were loosely based upon the idea of the "worker-peasant alliance" (see chapter 4). In accordance with the notion that social relations were at root harmonious and without underlying conflicts of interest, a contemporary view was that

> Each economic relation is two-sided. The cooperative must sell and pay [taxes] to the state, so the state must sell means of production, raw materials, and all the products that serve the lives of the peasants.
>
> With regard to politics; the cooperative is a collective socialist economic organization freely set up under the assistance of the state. Thus in each situation the state must help the cooperative. . . . The state is the property of the proletariat, the workers. The peasants are the friends and allies of the workers, so the proletariat must help the peasantry. (Hanoi I)

In keeping with both the CMEA thinking referred to above and the general idea that horizontal relations between base units were inappropriate, it was clear that a cooperative did not have the right to trade directly in merchandise with another cooperative. Only vertical relations with the relevant state trading organ were legitimate. But, in a typically pragmatic and idealistic Vietnamese manner, cooperatives could—if not should—help each other ("economic relations are not allowed, but relations of help are"[33]). Furthermore, these restrictions mainly applied only to current output. A cooperative could sell equipment "if it is their property, i.e., was bought with their own money." A cooperative could use another cooperative's labor, but could not sell land.[34] In practice, of course, such stipulations could easily provide useful arguments to justify semilegal activities.

In practice the Statute was widely ignored. It did, however, provide a guide to the rules of the game. It also clearly shows up certain basic characteristics of policy makers' thinking. The various elements of the cooperative were to work together, fully understanding the correctness of the party line and morally committed to socialist construction. The cooperators were bound to work for the collective, and in return should have received, not only the benefits of increased production, but also a relatively comprehensive system of social servies—income floors, education, health, and other items. The development path for production was relatively undefined. It encouraged diversification, consolidation of existing staples production, and an increase in labor productivity resulting from improved techniques without ending labor sharing. Except for residual tendencies toward "authoritarianism"—which were condemned—no fundamental conflicts between groups in the cooperatives were foreseen. The implicit solution to such conflicts could be found in the view taken of the cooperative as both a political and an economic unit, and the resulting concentration of powers in the local leadership's hands.

Rural Administration and Economic Management

The discussion must now move toward establishing an analytical framework. This will aim to answer basic questions of political economy, most especially those asking how economic agents acquire and dispose of economic resources. It is also necessary to establish precisely who or what those economic agents are. The next section therefore examines some possible general implications of the above system. The effects of the household economy are introduced later.

The Collective

The factors influencing access to economic resources in the cooperative were far from straightforward. This is true even when they are considered in isolation from the private sphere.[35] In the first instance, local cadres determined any cooperator's access to means of production and consumption. The management team[36] of the cooperative in principle controlled the cooperative's assets and the distribution of the cooperative's output to its various destinations. It was also the agency for distribution of consumer goods and means of production obtained from the state. In principle this conferred considerable control over economic resources:

—The cooperative's land was allocated by the Management Committee to the brigades whose leadership then organized its use. Part of this land was not given out to the brigades, but allocated to households for use as "5 percent" land—the "private plots"; this use was regulated.

—All other means of production belonging to the collective (livestock, tools, fertilizer. etc.) were also subject to the cooperative's management team.[37]

—Labor had the duty of working for the collective; the brigade leadership directly supervised collective work.

—Collective output was allocated by the cooperative's management team, who were responsible for ensuring an appropriate distribution among cooperators, state, and the various funds.

Control over economic resources provided a basis for the social position of members of the management team within the commune. The party's occupation of key positions in the management team therefore provided one basis for any social authority exercised by the party. But there was clear potential for frictions to arise within the management team between the Management Committee and the brigade leadership.

These frictions stemmed at root from fundamental differences between the sources of social and economic authority of the two groups. These should already be apparent from the discussion of the formal structure of the cooperative. The brigades both were responsible for and actually carried out collective production. This meant that the brigade leadership was the agency responsible for the resulting output and, in principle, delivered it to the Management Committee. It appropriated collective output in the name of the cooperative.[38] But the only channel for procurement and deliveries from the state was the Management Committee, which was the sole level allowed direct relations with higher levels. Thus in the sphere of distribution the brigade leadership and the Management Committee had different functions. This created scope for conflict. Fundamental aspects of a cooperative's structure created an opposition between the Management Committee and the brigade leadership. Put crudely, the former had to induce the latter to part with output that the latter directly controlled. A similar opposition arose from production conditions. Since the labor process was the responsibility of the brigade leadership the Management Committee had little direct contact with it. As a result, brigade leaders had a closer and more direct relationship with cooperators than with members of the Management Committee. The natural tendency was for this proximity to result in a divergence of interests between brigade cadres and those in the central organs of the cooperative.[39] This simple but fundamental opposition in the spheres of both production and distribution resulted in a wide range of overt conflicts discussed in detail in later chapters.

A great variety of factors influenced the way in which the management team as a whole exercised its ability to control economic resources. The effects of the household economy are considered in the next section. Among factors operating in relative isolation from the household economy, consider the following:

Personalities involved. Here two issues arise: first, the existence of individuals of strong character capable of providing a basis for alliances within the dual-level management system, or, on the contrary, of preventing such agreement; second, the direction in which dominant personalities wished to take the cooperative. This could reflect their prerevolutionary class background or be largely random and stochastic. Some cooperatives and brigades were lucky to have had dynamic leaders.

Influence of cadres from higher levels in the rural administration. This could act, for example, through their ability to control appointments (e.g., by stipulating that only certain individuals could be members of the cooperative's management team). Alternatively, local cadres may have responded to material and nonmaterial incentives (such as special party schools, promotion outside the commune, and emulation campaigns for "model" rural cadres). The Management Committee, and especially the cooperative manager, had direct relations with higher levels in the rural administration. It thus acted as an intermediary between brigade leaders and cooperators, on the one hand, and higher levels, on the other. This accentuated differences between brigade leaders and Management Committee. The latter had ways of securing direct access that the former lacked.

Influence of the free market. The analytical distinction can and should be made between the influences of the "free market" and of the household economy. The presence of market opportunities and the possibility of spending cash incomes so generated could have considerable influence. It could encourage embezzlement of collective property and other efforts to "privatize" collective assets so as to realize the differential between free market and state prices. If material incentives were appropriate, pressures could arise for the cooperative to be run for the sole benefit of the dominant members of the management team. State procurement, cooperative accumulation, and cooperator incomes would be minimized in favor of illegal sales on the free market.

Influence of cooperators. In the likely absence of effective democratic procedures, the cooperators' main channels of influence upon the management team were:

—The ability to call in higher levels (informally) to ensure appropriate behavior. This depended upon the efficacy and consequences of such action: would higher levels respond, and, if so, would it result in a net outcome favorable to the cooperators?

—Direct action or "social persuasion." The great population density in the villages placed limits upon the extent to which differences could arise between the general wishes of the cooperators and the management team. There were always possibilities for direct action, but the possibility that higher levels could, in extremis, retaliate physically limited what could be done. Complaints ("hourly") against the brigade leadership were likely to be both easier and more effective than those against Management Committee members. Again, the relative proximity of the brigade cadres to the cooperators would tend to give them a different set of perspectives and interests from cadres in the central organs of the cooperative.[40]

—The ability to act so that the cooperative's development was to the management team's direct benefit. For instance, cooperators could influence cadres by not sabotaging affairs so that the team appeared incompetent in the eyes of superior levels. Alternatively, they could allow selective implementation of development projects required of the cooperative but not to the cooperators' direct benefit (e.g., building a collective pigsty). Because Management Committee cadres were responsible to their superiors for the cooperative, this line of influence again acted differently at brigade and Management Committee levels.

Influence of kinship groups. Although rarely documented, the influence of relatives upon members of the management team could be strong.[41] Such factors would result in cleavages and alliances outside the system of rural administration and economic management. A manager with relatives in one particular hamlet would be suspected of favoritism; an official with many surviving nonworking elder relatives would be influenced in his attitude toward the social welfare fund of the cooperative; a childless official might be less concerned about the educational system.

Although this area points toward interesting avenues for research,[42] it is included under the general heading of the "private interests" of the management team. It was not likely to be systematically differentiated by levels of economic management. The most important consequence was probably the creation of a basis for differences between individuals and an enhanced potential for social alliances and cleavages.

Influence of the cooperative. As collective production rose and the level of incomes from the cooperative increased, so should the team's material well-being. If collective distribution to all was below subsistence, then officials too had to obtain extra supplies from somewhere. But it was not generally possible to identify the interests of the cooperative (e.g., expressed in terms of the level of collective income) with those of the management team.

Some Preliminary Conclusions

While still abstracting from the "household economy," the above considerations permit the drawing of some preliminary conclusions. It is possible to derive certain alternative patterns of surplus generation, mobilization, and use in the cooperative. These differed according to whether or not production conditions yielded a surplus above the level of subsistence required for labor reproduction.

Collective output was initially appropriated by the brigade leadership. Three groups of agents then competed to control it: cooperators, the two management levels of the cooperative, and outside agents. Production conditions—the need for seeds, feed, stocks—would influence the outcome, but the resulting pattern of distribution would ultimately both reflect and reinforce social pressures of the types examined above.

The consequences for the development of collective production largely depended upon the use made of collective product by the management team in its relations with cooperators. The amount allocated as collective distribution for work done, and especially whether it was capable of guaranteeing subsistence, would affect the quantity and quality of labor supplied. The method of its allocation (i.e., the use of penalties and bonuses) would also have an impact. But the team itself could seek to retain a share of output for a number of possible uses. Historical behavior encouraged the use of formal positions of authority for family gain. It also supported a view of them as the proper object of struggles for status and as sources of authority. The need to acquire support and the expected consequences of office suggested that the management team would use part of collective product in the following ways:

—Above-average household consumption—in conditions of absolute shortage, simple subsistence.

—Consumption designed to reinforce group solidarity and emphasis rankings within the dominant group. This varied from that focusing upon the entire cooperative (e.g., financing of social services such as education) to that focusing upon clique-formation and gaining better relations with superiors.[43] Examples of the latter would be feasts for cadres "coming down" from higher levels or weddings for children of high-ranking members of the management team.

The precise pattern of distribution that resulted would depend on the wider balance of social forces. For instance, the precise location of (illegal) sales of collective output onto the free market would be affected by the internal balance of power within the management team, and especially that between the brigades and the Management Committee.

It would be quite misleading to view the level of deliveries to the state as a residual. In the overall power balance cadres occupying superior positions in the state apparatus had considerable potential influence as possible providers of consumption goods and means of production. They also possessed a more general authority as members of regulatory and supervisory organs superior to the communal administration. But in situations where collective distribution was so low as to threaten cooperators' subsistence requirements, the issues involved would become qualitatively more important. The management team would then be under considerable pressure from cooperators for a change in the situation; the level of procurement and of collective economic activity itself would be under attack. A possible compromise would involve an increase in the permitted area of the private plots, which provokes wider questions discussed below.

In attempting to understand the issues involved it is important to recall the continuing importance attached, both in dynastic times and later, to the functions performed by local institutions in the economic sphere. This was associated with the fundamental structure based upon the management team's position as an intermediary between peasant-cooperators and higher levels of the centralized state. Of itself this was an important source of the local cadres' authority. In addition, it also led to the potential for division between the necessarily outward-facing Management Committee and the inward-facing brigade leadership. In the collectivized system, this corresponded to their different functions in the spheres of production and distribution. The institution of collective production meant that there was a radical difference between sources of authority under the old and new systems. As will be seen, a crucial problem presented by the cooperatives to policy makers was to ensure that the level with which the state had to have direct relations—the Management Committee—did not confront a situation where an excessive concentration of economic power in the hands of brigade leaders had undermined its position. If this happened, superior levels could not use the Management Committee as a mechanism for controlling the cooperative.

Private Household Economy

The Statute clearly and explicitly permitted own-account activity on the "5 percent" land. A household's use of its own net bundle of resources was, however, in principle constrained within parameters set by the cooperative. Within the household, status based upon sex and age reflected and reinforced a similar division of labor. A fundamentally patriarchal system, with an emphasis upon respect for age, persisted despite modifications resulting from, among other things, political campaigns and the operation of the cooperative system.[44]

In practice a variety of patterns of surplus generation and interactions between cooperator's own-account activities and other areas would tend to emerge. Any inadequacy in collective staples distribution that threatened physical reproduction would reinforce the basic dichotomy between own-account and collective activity. This dichotomy was not a simple opposition. In some instances a joint interest could support interactions that supported both types of activity. Livestock-rearing and certain forms of output distribution in practice tended to confirm this (see chapters 7 and 8). The key point is that opportunities for own-account activities based upon the private household economy and access to markets presented cooperators with economic alternatives. Thus a cooperator household would compare the values of economic resources used in own-account activities with the opportunities offered by the management team. Two issues together make the entire process extremely complicated. First, resource allocation decisions by cooperators were highly conditional. This was because the resources they controlled were themselves the outcome of an ongoing struggle to determine the content of the collectivized system. Property rights of economic agents within that system—cooperators, brigade cadres, the Management Committee—were all "immanent," and realizing them was not a foregone conclusion. Second, the value of those resources was also conditional and depended upon the decisions of other economic agents. Production was not solely for a market, where prices received could be assumed exogenous, but for demand within the collectivized system. Thus the "value of economic resources" available to a household in a fully collectivized system is by no means a simple category. It has to be understood in the context of the immanent and conditional property rights of that system as well as the nonmarket nature of many distributional relations within it.

In addition to direct marketing and own-consumption of private household production, a household could use surplus product to help the formation of social and political alliances. Such attempts to improve the household's social position could focus either upon members of the management team and other parts of the rural administration or upon other households. In addition, limited outlets for investment within the household existed in such areas as house-building and the orchards and ponds. If rice cultivation did occur in the household sector, the higher yields almost universally expected resulted from more intensive cultivation, improvements in techniques, and favored access to means of production available in the local household economy. Although modern inputs could be secured, this probably had to be done illegally. At the margin, work within the framework of the household economy would almost always appear more profitable in cash terms than work for the collective. But, as should already be clear, this views the decisions taken by cooperators too narrowly. Collective work generated values that would not be reflected in the direct remuneration. Such work was a duty, so not doing it could generate long-term social costs such as being marked as an antisocialist element. Support for the collective helped maintain it as a protective "umbrella" against any unwanted demands from the state. Collective institutions could help provide insurance against risk. Thus, while the immediate reward to own-account activity at the margin would usually have exceeded that from collective labor, this does not mean that cooperators would, on balance, wish to abandon collective work altogether. To understand their strategies, it is essential to appreciate the nature of interactions between the collective and the private household sectors.

Interactions Between Collective and Private Household Sectors

The Statute clearly stated the contingency of cooperators' private property rights within the cooperative, for they only had the right to "expand the minor household economy in conditions that do not interfere with the cooperative's management of labor." In practice, it is almost impossible to imagine a situation in which there would be no such "interference." Legitimate expansion of the household economy was therefore conditional upon the interests and powers of other agents—most importantly the cooperative's management team.

The Statute implied an ability to isolate the collective from the interests of the household sector that in fact would only be realized in two radically different and highly unlikely situations: first, when the cooperative's labor management was based upon straightforward coercion, so that economic comparisons of relative advantage by the cooperator and the cooperative were both irrelevant; and second, when, despite the absence of coercion, economic comparisons were irrelevant. The precise conditions for this to be so clearly depend upon such issues as a possible discontinuity in the rewards to collective labor (e.g., as a result of explicit labor duties that have to be filled before collective incomes are obtained), strong moral incentives (such as a commitment to the revolution), the nonavailability of certain goods in one sector (e.g., meat, with no trading oportunities), and so on. In essence these arguments require a "separability" of decisions about the two activities which, while feasible, is somewhat far-fetched.

Thus under normal conditions when neither of these two conditions held, the cooperator must be expected to adjust his or her activities so as to benefit as best as possible. Two crucial questions would then be—if collective incomes were below subsistence, where could increased supplies best be found? And, if collective incomes could guarantee subsistence, how would additional effort generate the highest net reward? If state supplies of consumer goods were limited, then the private household sector would presumably be preferred in that it more easily generated a marketable surplus.[45] The existence of social goods such as education and other collectively supplied facilities and the possibility of selling suprasubsistence collective income on the free market[46] would mitigate the effects of such shortages on the collective sector. In such circumstances the incentive effects to participation in the cooperative would exist side by side with other factors such as the impact of penalty-bonus schemes on "free-riders."

Further relationships between these two sectors arose from the management team's position as the controller, in principle, of the cooperative's assets. "Privatization" of these assets by allocating them to the private household sector was one possible development. Subject to the regulation of higher levels, who may nevertheless have been willing to tolerate it, this strikingly recalls the possible attitudes of the dynastic state toward communal land in the traditional commune. Such phenomena may be differentiated as follows:

—Use of such assets to expand the private sector activities of members of the management team: either a simple increase in the area of their own family plot, or some form of tenancy arrangement with a cooperator or an outsider.

—Use of such assets to help meet cooperative goals. It may have been thought possible to encourage cooperator participation in this way: for instance, by making holdings of such land conditional upon fulfillment of labor duties, or the paying of a "rent" set at a level designed to help develop collective production.

—Use of such assets to acquire support within the commune. If adequate support could not be gained by the rural authorities through the "ordinary" operation of the cooperative, then an increase (perhaps to selected groups) in the area of cooperative land farmed in the private sector may have been an alternative. Clear pressures for this would exist if collective incomes were well below subsistence and perceptions of possible increases pessimistic. A wide variety of outcomes would depend upon such conditions as marketing opportunities, the internal power balance, and the pressure upon subsistence.

An important basis for the social position of the rural administration was the cooperative. Because of this, as usufruct to such encroachments became entrenched so the position of the cooperative per se in the political balances involved would change and probably diminish. Certain of the distributional functions of the Management Committee (e.g., procurement deliveries) would inevitably persist while collective production declined in importance. At the same time the role of the brigade leaders would fall away steeply as cooperative land became increasingly privatized. The formal functions of the Management

Committee would essentially reduce to indirectly representing the state in the collection of taxes. The management team, perhaps still receiving a "rent" distinct from the state's taxes, would then preside over a cooperative whose assets were let out to various households or individuals. Such a cooperative would have an only "nominal" existence.

A further area of interaction between the two sectors would occur if factors of production other than labor and land moved from the household sector to the cooperative. Possibly related to labor allocation, similar arguments apply because in most instances there existed an opportunity cost to the cooperator for such factors of production if (as is most probable) they could have been used in the private sector. Supplies to the cooperative of inputs such as natural fertilizer and piglets would be clear examples of such interactions.

The implications of the above considerations for any empirical work are immense. The conditions under which interactions between the two sectors can be ignored are highly unrealistic. It follows that any analysis has to take into account not only the alternative opportunities faced by cooperators but also the possible dynamic effects of their strategies upon the collectivized system. If extensive privatization of cooperative assets has taken place, then extreme care must be taken to penetrate sufficiently deeply into rural society in order to find out precisely who or what is controlling access to economic resources. If such phenomena were present, then there will be considerable incentives to misinformation and deception, especially of regulatory agencies and higher levels.[47]

The incentives toward bias in reports would reflect the balance of forces acting upon local officials. The desire to keep outsiders out of the commune would encourage reports that confirmed the prejudices of higher levels: policies would be reported as implemented when they had not been. The true structure of relations in the cooperatives would not, therefore, easily be known to higher levels. The precise effects upon output data would clearly depend upon the balance between the likely costs (e.g., the higher taxes likely to result from higher declared output levels) and benefits (e.g., the continued independence from investigation moderate success would confer).

The above suggests that it is absolutely necessary to reconsider the dominant position usually attributed to the cooperative in analyzing collectivized agriculture. If authority is concentrated at brigade level, or if large-scale privatization of cooperative assets occurs, or if the multiple functions adopted by rural administration preclude a clear division between economic and noneconomic factors, then the cooperative is only one part of those elements of sociopolitical organization that determine economic resource allocation, production development, and surplus mobilization. The next chapter argues that to the extent to which it is useful to define a "basic unit" of cooperativized agriculture, all these elements should be included so as to permit analysis of the various economic and social roles that the cooperative's management structures could play within this larger unit.

3

A Reexamination of the Basic System

The basic unit of analysis used here includes the cooperative's two economic management levels as well as the household. These three are treated as a triad called, for want of a better term, a "collectivity." This unit is understood conceptually to be quite different from the cooperative. The basic concept used to understand interrelationships within the collectivity among these three levels is that of dominance, in the sense of one level's ability to determine the outcome of activities carried out by the other(s). Use of this concept allows consideration of the great variety of factors, especially the strictly noneconomic, that influenced economic decisions. The term is used to illustrate certain systematic patterns of relations among the three levels of the collectivity. These different patterns accompanied systematic variation in the development of production and surplus mobilization, and they tended to produce similar outcomes despite having different origins.

Production and Output Control as Sources of Authority, and the Immanence of Property Rights

The extent to which formal property rights were realized[1] reflected the "dominance" of one level by another. If a level formally entitled to certain property rights could not in practice realize them, then that level was dominated by others.

Thus the persistent problem of the lack of effective control exercised by the "cooperative" over its "own" resources and economic activity within it appeared paradoxical unless and until it was interpreted to mean that the "wrong" levels were dominant. Any attempt by the authorities to rectify this situation would entail establishing the cooperative Management Committee's statutory rights. For example, this could mean ending effective control over land by the brigade leadership or the cooperators and instead instituting control by the Management Committee "in the name of the cooperative."

The basic distinction made here between different types of collectivity is therefore at root quite straightforward. These differences are a simple reflection of the extent to which a given cooperative operated according to the policy prescriptions of the central authorities. If such policies are implemented, then the property and other economic rights of the cooperative are fully realized. The Management Committee of the cooperative then dominates the collectivity. The cooperative functions according to the norms of the Statute and—in the period 1974–79—closely follows the NMS in its management methods.

The nature of the control that resulted from and reflected the dominance of one level by another can also be illustrated by means of the distinction between control over production and control over output distribution. When technology is relatively static, attempts to develop production may tend to generate conflict over distribution more frequently than over production; the first priority would be to control resources for financing development. As technology changes, so new ideas and knowledge would usually be required to adapt production to the new methods available. If so, control over production methods will change hands unless old leaders can learn the new methods. This gives scope to an increasing role for "technocrats" in economic management and, therefore, in social life. But such a "technocratic" position would differ from one based upon control of distribution. In a situation where some group or groups have an essentially distributive function (for instance in ensuring that taxes are paid), this, while conveying authority, has little in common with the technocratic position. The emphasis will be upon ensuring a "proper" distribution of output, and, within production units, upon appropriation. Supervision of the labor process becomes less important than ensuring that output goes to its proper destination.

A "modernizing" group, eager to use resources to finance new methods, is at first seen as simply another claimant on existing output to be resisted in the same way as a new tax collector or rising local leader. Conflicts focus upon distribution. But if the group is successful it will extend its influence deep into the production process, increasing labor supervision and other controls in order to enforce the use of better techniques. This inevitably tends to reinforce control over output, via the creation and use of an accounting system, and so forth, and so the two sources of authority—control over production and control over output distribution—both develop with the development of production. They can, however, be clearly distinguished, and the social dominance of a group concerned to

Table 3.1

Agricultural Producer Cooperatives' Management Structures and Balance Between Levels

Characteristic	Nominal	Active	Model
Development of collective production	Poor	Good	Good
Management Committee–brigade relations	Brigades dominant	Apparent compromise, great variations	Management Committee dominant
Distribution system	Brigade-level	Varied, usually coop-level	Coop-level
Development of sub-subsidiary branches	Weak, de-centralized	Sometimes very strong, decentralized	Strong, centralized
Development of collective pigs	Minimal	100–500, unprofit-able, problematic	More than 1,000
Functionally special-ized brigades	Never	Limited	Extensive

improve methods of production implies a shift in the relative importance of the two determinants of the economic origins of such dominance.

The use of the "collectivity" concept permits greater precision in the use of the words "cooperative," "collective," and "household" (or "private"). The first now refers to actions taken in the name of the cooperative by the Management Committee, the second to activity outside the household controlled by the brigade or committee, and the third to activity inside the household and primarily controlled by it.

Nominal, Model, and Active Cooperatives

A typology of the different social relations within the delta cooperatives under study can now be constructed in terms of the relative dominance of each of the three levels discussed above. This is summarized in table 3.1.

In "nominal" cooperatives the focus of economic power in the collective sphere lay at the brigade level. In "active" or "model" cooperatives (grouped together under the heading "advanced") there was some integration of economic activity at the level of the cooperative. In model cooperatives this was carried to an extreme and economic power concentrated almost exclusively at the level of the Management Committee. In active cooperatives a decentralization of power reflected a more diffuse pattern of dominance.

In nominal cooperatives (prereform My Tho is a well-documented example—

see chapter 6) the Management Committee had no control over the production process; it acted primarily as a conduit for product flows both inward and outward and was subordinate to the brigades, which controlled production, distribution, labor use, and remuneration. A minor portion of product may have been allocated by the Management Committee in order to pay pensions and so forth. In such circumstances the "cooperative" provided a useful way of keeping outside authorities at a distance, for because the brigades had land, labor, and means of production and delivered a share of "their" product to the state via the Management Committee, the state did not deal with the level that had real power. Intead it had to cope with a "front," which, according to the Statute, had responsibility for dealing with the state but could not be used to dominate the cooperative because it had no great influence.

A model cooperative was one where the Management Committee dominated the brigade leadership. An indication of the strength of the forces supporting a nominal structure is that the creation of a model cooperative often appeared to require outside influence—usually in the form of a cadre team backed up by extra supplies of modern inputs and higher-level pressures. Such outside intervention tended to disturb the status quo in a way that allowed the Management Committee to establish control over both collective output and production; subsequently, as production developed, it could reinforce its position by successfully and effectively regulating both collective and household production at the level of the cooperative. This would constitute a process of "reform of the management system." The extension of control over methods of production into the domain of the brigades was both strengthened and facilitated by creating specialized brigades both within the cultivation branch and in the other branches that used or produced means of production (e.g., jute-using mats; manure-generating pigs). Control over output was reinforced by the accounting system and the generation of information about the process of production (labor outlays, etc.). This helped to prevent expropriation of what could now properly and accurately be called "cooperative" property by aiding the formulation and implementation of regulations. Such systems often had a major positive effect on the development of production though allowance must be made for priority resource allocation to showpieces.

In a position intermediate between model and nominal cooperatives come those of the active type, where the Management Committee did not dominate production in the same way as in the model cooperatives. Its position was far stronger than in the nominal cooperative, but the weight of the household economy and artisanal production was greater because the brigades retained some independence and there was no drive for centralized managerial control over all areas. Corresponding to a relative lack of any clear division of authority between the two management levels was a greater unity within management; the interplay of internal factors was more important, and there may have been no team of outside cadres. Such cooperatives nevertheless revealed a scope for more "balanced'' development, and (to anticipate chapter 10) imply that the complex centralized management system of the model cooperatives was often unnecessarily complex and unwieldy. The hypothesis is that the complexity of the model system prescribed by the party was a control technique intended to reinforce the position of groups that had little other than state support for attempts to reform nominal cooperatives.

Corresponding to these underlying differences were certain basic issues governing control over means of production and labor. The distribution system could be based on the brigade[2] or the cooperative level, depending on the balance of power. This in turn had implications for the control of labor, since in the absence of other factors a Management Committee without control over distribution could hardly control labor. Quite apart from this, there was the question of land and other means of production. The prevalence of brigade ''ownership'' of land was confirmed by the widespread problem of ''land-scattering,''[3] where the land worked by a brigade was that brought in by its members when they joined the cooperative; these parcels were scattered throughout the commune because Land Reform was carried out at village or commune level, whereas the earliest cooperatives and production brigades were based upon hamlets. ''Rationalization'' of landholdings required enforcement of the cooperative's (i.e., the Management Committee's) rights to land—but this was seen as ''taking one brigade's land to give to another'' and led to conflict. These were the two major areas of dispute, but similar problems were encountered elsewhere: the basic issue remained one of control and authority.

Implications for the Development of Production and Surplus Mobilization

The above typology had implications for the patterns of surplus generation, mobilization, and use as well as for the development of production.

Model Cooperatives

In model cooperatives, the concentration of control over economic activity within the collectivity in the hands of a dominant Management Committee corresponded closely to the prescriptions of the Statutes. Surplus product was generated by brigades, mobilized out of them by the brigade leadership, and then used for a variety of ends: the development of production, the consumption needs of cooperators, and the requirements of the state. Success with these permitted further advance as cooperators responded to material incentives and sales to the state financed purchase of means of production. Specific to the model cooperative, however, was the cooperative-wide control over economic activity, which had five important dimensions.

First, diversification of cooperative production into the production of means of production: certain areas of activity (e.g., livestock) generated inputs to other areas (e.g., manure), so involving an integration of production at the cooperative level. This was quite different from the impact of diversification into production of other goods, which did not usually entail integration at this level.

Second, direct accumulation through the investment of the cooperative's own resources (primarily labor) in various projects: this also tended to raise the level of integration of production, because access to such assets and the services they provided were regulated at the level of the cooperative.

Third, indirect accumulation through the mobilization of exports outside the cooperative and the creation of finance for the purchase of means of production. By effectively reducing the cooperative's self-sufficiency, this dimension deemphasized the ''self-reliance'' that was a key element of prescribed policy. But the position of the Management Committee as the conduit for all trade by the cooperative inevitably meant that policies of indirect accumulation tended to reinforce the committee's position.

Fourth, improvements in production techniques introduced through cooperative diffusion of information about ''better practice'' methods. These reinforced the ability, resulting from the effective control over output and penalty/bonus schemes, to secure effective labor discipline.

Fifth, strict regulation of the private ''household'' economy. This tended to have strong positive effects on the supply of labor and means of production to the collective sector, at root by increasing the perceived relative net cost of participation in ''own-account'' activities.

A consequence for the overall pattern of development in the cooperative was the likely emphasis upon increased exchange with the state rather than autarkic[4] ''self-reliance,'' because the level directly fulfilling this function was now in control of the cooperative. The state was the only outlet for surplus mobilization outside the borders of the ''collectivity'' because of the strict regulation of others by the Management Committee.

Nominal Cooperatives

In nominal cooperatives, concentration of control over collective economic activity in the hands of the brigade leadership meant that the cooperative had a merely nominal existence, performing certain intermediary functions. This was radically different from the prescriptions of the Statutes. The relative proximity of cooperators to the dominant level could increase their influence or, on the contrary, subordinate them to the interests of the brigade leadership. This eroded the cooperative's role as a controller of economic activity with effects that were contingent upon the development of the macroeconomy and the demographic pressure upon subsistence, for instance:

—privatization of cooperative assets;
—sales onto the free market from a variety of sources;

—accommodation to the requirements of the household sector resulting in limited labor mobilization, poor labor discipline, and a low quality of work in the collective sector, coupled with a weak regulation of household activities;

—minimization of deliveries to the state when the state had little to offer in exchange;

—no production diversification within the cooperative, and above all no development of specialized branches of production involving an integration of production at levels above the brigade;

—no direct accumulation that involved communal or village-level infrastructural investment (such as water-works) where this might entail a shift in the level of control;

—limited indirect accumulation on account of low deliveries to the state;

—an accumulation constraint on diversification into new areas: thus the brigades' limited initial conditions may well have hindered development in the brigades that, if rice monoculturalists, would tend to remain so.

Active Cooperatives

In the intermediate case of the active cooperative, a balance between the two management levels tended to reflect the lack of clear dominance. Any integration of production at the level of the cooperative coexisted with phenomena more closely associated with the nominal cooperative. A key feature of the active cooperative was therefore the pattern of diversification of production: in the absence of control centralized in the hands of the Management Committee, such diversification would tend to perpetuate brigade independence, and so place limits on any diversification of cooperative production into the production of means of production.

In practice, an active cooperative's direct accumulation potential, although clearly much greater than for a nominal cooperative, depended greatly upon its ability to mobilize labor. Similarly, the possibilities for indirect accumulation depended upon the precise nature of opportunities for exchange with the state.

Conclusions

This cooperative typology implied the set of characteristics summarized in table 3.2. A cooperative's transition from one type to another, or the predominance of one type at a national level, depended upon a number of exogenous factors, notably the quality of the local leadership, the availability of resources from the state, the relative attractiveness to cooperators of collective activity, demographic pressures, and the form of influence from superior levels: especially the possible presence of a team of outside cadres.

The different cooperative types implied different patterns of production diversification. This was associated with the implications of such diversification for

Table 3.2

Summary Table of Predictions

Characteristic	Nominal	Active	Model
Diversification			
Means of production	Nil	Limited	Extensive
Cons. goods	Limited (unless preexistent)	Extensive	Extensive
Direct accumulation	Very limited	Contingent (up- on cooperators)	Extensive
Indirect accumulation	Very limited	Contingent (upon state)	Extensive
Technical improvements in production	Very limited	Contingent	Extensive
Regulation of household economy	Weak	Weak	Strong
Privatization of coop. assets	Strong	Weak	Weak
Sales onto free market	High	Contingent	Low
Sales/deliveries to state	Minimal	Contingent	Excellent

the local power balance. In model cooperatives, diversification of production into production of means of production outside the rice-producing brigades much reduced the brigades' independence. The ''general'' brigades had then to secure supplies from outside their direct area of activity and therefore exposed themselves to monitoring and control by the cooperative's Management Committee. The existing power structure would try to limit diversification into the production of means of production because of the importance attached to brigade independence and the expected adverse effects on it of such a need to obtain inputs from elsewhere in the cooperative. By attacking rice monocultural brigades' autarky in a nominal cooperative, such changes therefore disturbed the balance of local interests.

In active cooperatives where local interests favored production development (if not entirely along the ideal lines of a model cooperative), the attitudes toward such changes were likely to be quite different. The relative balance between local interests meant that there was no need to attack brigade autarky in order to assist the takeover of the cooperative's structures by reformist progressive elements.

Part II

North Vietnamese Agrarian Policy, 1974–1979

4

The New Management System and the Thai Binh Conference

This chapter presents the main features of the prescribed management system for cooperatives in the mid–1970s—the NMS. This was formalized in the tract "Toward a Large-Scale Socialist Agriculture" (1975) by Le Duan, party general secretary. Next, two famous model cooperatives that applied this system in practice are examined. Finally, the thinking behind this policy and the cooperativization movement as a whole are considered. It is then possible to reexamine the NMS in the context of the wider agrarian problems facing the DRV/SRV leadership. These were closely bound up with the general issues of policy nonimplementation and cooperative nominalization that made up the Agrarian Question. It is therefore essential to appreciate the nature of the NMS before moving on to discussion of the problems it was intended to solve. The NMS was a powerful set of management methods prescribed for everyday use in the cooperatives. Its failure was a failure of implementation at the grass-roots level, and the Agrarian Question of the 1970s has to be seen in these terms. It is therefore misleading to present the theory of the NMS before explaining how it worked in practice.

The New Management System—Basic Principles

The New Management System[1] was based on three principles: unification of economic activity, of management, and of distribution.[2] This triple unification was to occur at the level of the cooperative, and the cooperative's plan and control system were the key mechanisms. Each Management Team was to prepare a production and distribution plan for approval by the district. These local plans were then meant to become part of the DRV's economic plan.

Implementation of the cooperative's plan presupposed some operational management structure at the level of the cooperative. This would allocate resources to the cooperative's various subunits—brigades and teams. These subunits were to specialize by function and by branch. The Management Committee's main means of controlling economic activity was a system of contracts[3] with the brigades. For each brigade, these specified both the vector of inputs (including labor) and the corresponding vector of outputs. A system of labor norms was used to calculate unit labor requirements for each constituent task required for the prescribed output vector. This gave a way of calculating the detailed labor requirements for the output plan. The brigade's workpoint allocation for the plan could then be calculated by attributing values to the work. Payments could vary according to task and worker. These payments to brigades were also the basis for calculating labor incomes. The cooperative was to select the workpoint valuations for each brigade so as to achieve "unified distribution."[4]

No possible contradiction was admitted between the interests of the cooperatives, the cooperators, and the state.[5] After meeting tax obligations, the NMS was supposed to increase both collective distribution and deliveries to state purchasing bodies. In addition, it was also supposed to exploit local conditions through output diversification into industrial crops, other exportables, and means of production—for instance by pig-rearing. Pig meat could be exported, and the manure used as an input to the cultivation branch collective. Pig-rearing therefore became a key element of the strategy irrespective of the strength of the rationale for collective rather than private activity (see chapter 8).

In some formulations the production integration sought by the NMS was to occur at the level above the cooperative, that is, the district.[6] This level was to accelerate direct accumulation and the division of labor between cooperatives by coordinating their activities. Indeed, Le Duan had even argued that

> It is not possible to establish [a balance in the division of labor between cultivation and livestock rearing] within a cooperative. . . . We must first start from the economic and technical plan for a whole district, [because] if all the factors of production . . . are "hermetically confined" within the limits of a cooperative, this will simply be an autarkic economy under another form.[7]

In practice, however, the district was of little direct relevance to the problems encountered in implementing the NMS. The focus remained the individual cooperative.

The essence of this managerial system can be described diagrammatically (table 4.1). Note that private plots were regarded as minor and peripheral. This was

Table 4.1

Schematic Outline of the New Management System

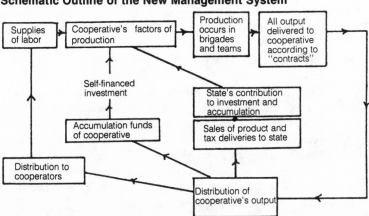

consistent with the view that the effective supply of labor to the production process was determined within the collective sector. Both in practice and in theory there were good reasons to suppose, on the contrary, that a cooperative's real labor supply would be highly dependent upon the relative opportunities in "own-account" activities.

Viewed in isolation from the history of the cooperative movement in North Vietnam, this simple system might appear to provide a rational solution to the problem of rural development. Shortages of industrial inputs, especially within the state sector, necessitated a self-reliant approach to the mobilization of resources for the accelerated generation of output and marketable surpluses. Once implemented, the NMS promised much: centralization of control in the hands of the Management Committee allowed for direct accumulation within the cooperative by using "surplus" labor. Central control also permitted production diversification into exportables and rapid technical improvement. But in practice the NMS was only rarely implemented, and its full relevance should therefore be reassessed. Before doing so, however, its application in model cooperatives must be considered in order to bring out fully the complexity and power of the system. These cooperatives were intended as demonstrations to rural cadres of the possibilities and practicalities of the system.

The NMS in Operation—Two Model Cooperatives

The two model cooperatives described here (Vu Thang, Kien Xuong district, Thai Binh; Dinh Cong, Thien Yen district, Thanh Hoa) were used both for emulation campaigns and academic discussion. Vu Thang was probably more important as a model for emulation purposes. The two together, however, were extremely well-known. As such, they represent the way in which it was intended that cooperatives implement the NMS. For other delta cooperatives, a major issue was the residual problems of localism resulting from the process of amalgamation. The two models had therefore to show how they had grown.

Recent History and Reasons for Amalgamation

Both cooperatives had reached the level of the commune comparatively early. They had both been formed from smaller cooperatives based upon constituent elements of their communes. Vu Thang had resulted from the amalgamation of four village-based cooperatives in 1965, while Dinh Cong came from two smaller cooperatives in 1969 after a lengthy previous history of amalgamations. From an initial total of eleven cooperatives and eighty brigades in 1959–1964, Dinh Cong had been set up in 1969 with eighteen brigades. In 1974 reforms reduced the number of brigades to ten, including specialized brigades.[8]

Amalgamation of cooperatives in this way was seen as important for both streamlining management and concentrating resources: "Both by amalgamating smaller cooperatives and by concentrating labor power and capital Dinh Cong had constructed for itself a material-technical base according to the way [i.e., according to prescribed policy] of moving from small-scale production to large-scale socialist production."[9]

In both cooperatives the small size of the earlier cooperatives had been seen as a constraint on diversification into production of means of production. In Dinh Cong the limited diversification of production was regarded as unsatisfactory: livestock-rearing and the brick and tile kilns had not been given "proper attention." But the smaller cooperatives had been growing a wide range of crops— rice, maize, and such readily exportable lines as cotton, peanuts, sugarcane, sesame, and beans, which were replaced after 1969 by pigs and kilns: "Dinh Cong then [1969] fixed the specialization of the cooperatives as the production of rice allied with the livestock-rearing branch [primarily pigs] and the correct development of brick and tile production."[10] Thus in 1969 Dinh Cong's Management Committee already had sufficient control to carry out substantial changes in the pattern of resource allocation within the collectivity. It was not confronted with a nominal cooperative.

The population density per cultivable hectare in Dinh Cong (ten) was more favorable than in Vu Thang (thirteen), where there had been considerable problems in providing levels of distribution sufficient for subsistence. Both cooperatives had populations near 3,800.[11] Vu Thang's land was of poor quality: immediately after cooperativization yields were unstable, averaging 2.0–2.1 tonnes of paddy per hectare. In some years the state had had to reduce taxes and provide assistance of 80–90 tonnes of rice. Despite early progress in amalgamation (four village-level cooperatives were set up in 1961), collective distribution never

Table 4.2

Labor Allocation in Dinh Cong and Vu Thang

	Dinh Cong			Vu Thang		
Percent labor in:	1970	1974	1976	1965	1970	1977
Cultivation	73	46	46	97	67	60
Livestock	5	12	10	1	9	10
Subsidiary branches	22	41	45	1	24	29

Source: Le Trong (1978); Nguyen Manh Huan (1979).

exceeded 8 kg per capita per month in the early 1960s. It had fallen to only 3 kg in 1964 when 40 percent of households wanted to leave the cooperatives. Eventually Vu Thang, like Dinh Cong, was able to divert resources to the collective pig herd, which received substantial supplies of paddy from the high levels of output attained in the mid–1970s. There was also a rapid growth in subsidiary branch activities that produced goods for export outside the cooperative.

Staples Production

By the mid–1970s the NMS had achieved impressive results in these model cooperatives in terms of the changing structure of output, absorption of modern inputs, and improved labor productivity. The power it conferred upon their Management Committees was, it appeared, the main reason for this: centralization seemed to work.

Dinh Cong's cultivation branch was able to increase gross output by 130 percent during the period 1970–76 while reducing its labor force from 794 to 524. Labor workdays per tonne of paddy also fell from 160 to 146 over the same period. Staples output increases were outstanding in both cooperatives. In Dinh Cong the lack of any mention of early shortages suggests that paddy output in the later 1960s exceeded gross subsistence requirements (approximately 600 tonnes). On this basis paddy output more than doubled in both communes, rising to levels around 2,000–2,100 tonnes in 1976–77. In the mid–1960s Vu Thang's reported output had been around 1,000–1,100 tonnes.

The available data also show a striking change in the fundamental orientation of the collective economy which moved decisively away from brigade-based rice monocultivation. Table 4.2 shows how the proportion of labor occupied in the cultivation branch declined rapidly in both cooperatives.

In these cooperatives pig manure from pig-rearing had been used to improve cultivation yields. In this they closely followed official policy on the role of the collective pig herd. In many other cooperatives manure was often diverted into producing feed for livestock.[12]

Priority resource access for such model cooperatives undoubtedly facilitated the rise in rice yields. Figures for supply of chemical fertilizers are not available, but the machine stocks were substantial. Dinh Cong possessed a total of twenty-eight assorted pumps, motors, threshers, mills, tractors, and feed preparers in 1976. Vu Thang had twenty-five.

Substantial efforts had been made to improve land, and this was particularly important in Vu Thang. Here access to machines for land-leveling was an important factor. A comment from a cadre of an adjoining commune explained Vu Thang's success succinctly—"They got the machines."[13] Naturally enough, however, these rapid increases in yields and mobilized surplus were not attributed to favorable resource endowments but to implementation of the prescribed management system. Even in these two models, however, there were slight differences.

Management Systems

Dinh Cong. An extensive quote concerning Dinh Cong illustrates the perceived relationship between product diversification and the brigade/team structure:

> The history of the development of the major forms of labor organization used in Dinh Cong is the process of the appearance and resolution of contradictions in the technico-economic sphere in order to ceaselessly reinforce and perfect the form of the fixed production brigades. From the initial fixed production brigades which only produced rice, . . . using neither labor power nor means of production rationally . . . the cooperative moved on to using mixed fixed production brigades, where each brigade produced a large variety of crops. . . . [But here] it was impossible to introduce new technical improvements into cultivation, and also there was no specialization of labor, so that it was at the same time impossible to increase crop yields and labor productivity . . . [then] . . . in 1965–67 the cooperatives . . . developed pig-rearing in the cultivation brigades and changed the mixed fixed production brigades into general fixed production brigades [cultivating plus livestock-rearing]. This in turn led to contradictions that arose day by day . . . from 1969 until the present when the cooperative has changed the ordinary fixed brigades into brigades specialized according to product. . . . And this is the main direction of development in the coming period for agricultural cooperatives. This form has overcome the large part of the contradictions of earlier forms. Nevertheless if the historical or production conditions allow, then in a cooperative it may be possible to use such forms in a rational combination.[14]

Here the evolving drive toward creation of a cooperative-level management system with the effective power to permit product diversification is clear. The initial mixed fixed brigades diversified into both staples and exportables. While the later addition of pig-rearing integrated production at brigade level, it worked badly in practice. The system introduced in 1969 established brigades specialized by branch and product while collectivized pig-rearing was centralized. The next issue concerned the viability of further brigade specialization within rice production. The main question here was whether to set up a separate brigade to provide ploughing services, an example of the production of means of production by a subunit of the cooperative. The final sentence quoted above was aimed at those keen to reduce brigade independence still further, and in effect replace them by a system of direct management similar to that of a state farm. The author thus took a "rationalist" position and went on to criticize those who set up ploughing brigades "as if there were machines" when they were still using water buffalo.[15]

It would therefore probably be unfair to dismiss Dinh Cong as simply a subsidized mirror of official policy. Nevertheless, the use of such a model as an example for all cooperatives reveals much about the wider problems faced by the collectivized system.

By around 1977 Dinh Cong possessed a complicated managerial structure, which is shown in table 4.3.

There was no reason to suppose that any part of this system was only nominal. Note the much reduced role of the general rice-producing brigades, who now no longer even controlled their seed supplies, aided by the powerful accounting system. Both corresponded to a clear dominance of the rice brigades by the Management Committee, who effectively controlled both production and distribution in the cooperative. Here economic activity was subject, as intended, to strong centralized authority aiming at unified control.

The diversification and integration of production implied by the above organizational layout was an important element of the NMS. This brings out one implicit target of the NMS: dominance of the brigades, and, through them, of the cooperators.

In cultivation the NMS abolished the semi-autarkic rice-producing brigades of a nominal cooperative. They were replaced by "fixed" brigades, who had to obtain essential means of production from elsewhere in the cooperative. The Management Committee arranged for supplies from the seeds brigade, the irrigation team, the manure team in the livestock branch, and for transport services when needed from the transport brigade. There was a thorough system of labor management.

When preparing the land for rice transplanting, each fixed brigade set up a number of groups and took a ploughman from each of the households looking after a draft animal. This system was temporarily modified in 1974–75 when they experimented with five specialized ploughing/harrowing brigades. These were abandoned after ten days, ostensibly because the productivity of the ploughmen declined.[16] For the job of planting rice seedlings, each fixed brigade divided up into small groups for transporting, transplanting, and uprooting "resting upon the work norms of the cooperative." Each worker was given targets by number

Table 4.3

Organizational Structure—Dinh Cong (ca. 1977)

Fifteen brigades and one team had direct contact with the Management Committee

A. Cultivation branch

Nine fixed rice-producing brigades

    Harvesting: division by task into reaping and bringing in, and threshing. Use of workers from outside the brigade to form group responsible for bringing in.

    Ploughing and harrowing: no specialized brigades. Division of brigade into groups based upon ploughman/animal combination.

    Transplanting: functional division followed by individual responsibility, closely monitored (allocation of specific item of work to each worker).

    Weeding, etc.: groups or individuals allocated work when necessary.

    Use of brigade ledger-clerks (one to every two brigades).

One irrigation brigade responsible for upkeep of canals (23 men).

One seeds brigade responsible for strains, planting, and sprouting.

One irrigation team responsible for supply of water to individual fields (12 men).

Total number of workers: 524

B: Livestock branch

One livestock brigade

    Pigs: vegetable-growing team, vegetable-processing team, manure team, team responsible for caring for pigs.

    Other animals: four teams responsible for fish, ducks, beef cattle, and goats.

Total number of workers: 111

C: Subsidiary branch

One brigade making bricks and tiles

    Two teams responsible for firing, four for covering tiles, two for brick molding, one for transport, one for general work.

One transport brigade, based upon units of cart/animal/pairs of workers.

One water-transport brigade, made up of sailing ships working according to contracts between state and cooperative.

Total number of workers: unknown, but 86 in bricks and tiles brigade.

---

and quality, and a specific area of ground to cover.[17] This allowed the cooperative to use competitions and enforce collective responsibility, and it was also seen as an effective mechanism for ensuring labor discipline—the plan for each day tended to be exceeded. The time required for transplanting was cut from forty days in the years before 1973 to fifteen to eighteen days. The time taken to bring in the harvest also fell, from forty-five days to less than twenty days. This was attributed to similar methods of labor management and an effective mobilization of students and school children.

Labor discipline of this quality was accompanied by close supervision: "Every day the brigade allocates work concretely to each group or individual according to the figures in the contract. At the same time the brigade head checks up on the groups out in the fields" (p. 58). In the early 1960s, however,

> each day, the brigade head would gather the workers together in order to lead them to work. At the end of the day, he would value the work and mark it down for each worker. After some time one became aware that there was a passivity in the way the cooperators worked in the brigades, that they were "leaning" on each other, arriving late and going home early, and that especially the system of responsibility for production and material rights were not closely linked, leading to a poor quality of work and low productivity, and ending in general damage to the collective economy. (pp. 55, 56)

Such quotations speak for themselves, showing the power of the NMS and the ability it conferred to ensure effective labor participation.

The operation of the NMS can be observed elsewhere in the collective economy. Livestock-rearing had developed along the prescribed path of concentration after the great difficulties experienced with the brigade-based system used in the late 1960s, when a number of small sties of fifteen to twenty pigs had been worked by local cooperators in turn: "This system [had] led to a general state of drift in the system of responsibility during work in general and livestock-rearing in particular, where it was impossible to introduce technical standards" (p. 59).

A key question was the reason for the earlier management's lack of effective control, but none was given. Before 1970, 100 kg of pork (liveweight) were losing 380 dong and needed 310 workdays to produce. In 1970 the cooperative provided "appropriate resources" for a single brigade in the branch and output subsequently grew rapidly. The collective pig herd rose from 525 head in 1970 to over 1,200 head in 1976. As in the cultivation branch, the NMS seemed to show that rapid increases in both output and productivity were possible. Although detailed data are not available, an examination of the economic viability of this branch will be carried out below for Vu Thang; data on Dinh Cong were insufficient.

Outside agriculture, bricks and tiles were Dinh Cong's main specialization. Until 1973 production had only been for use within the cooperative. Demand was high, for before 1954 there had only been six houses made of brick. Earth for the kilns was found in the area. When the cooperative was set up a fixed brigade replaced the ad hoc temporary brigades and teams that had been set up when necessary by the smaller cooperatives. The 1974 reforms that reorganized the cultivation branch left this system unchanged, but the internal organization of the branch was streamlined, with changes in the number of kilns accompanying the introduction of machinery. Prior to 1974 a number of contracted teams, with one

Table 4.4

Organizational Structure, Vu Thang

1964   34 fixed rice-producing brigades with an average of 8.5 hectares and 40 workers.

1965   Livestock-rearing brigades set up (i.e., the general brigades used in Dinh Cong were never used).

1968   Seeds brigade and basic construction brigade set up.

1970   Reorganization of the rice-producing brigades to around 15. Ploughing and harrowing brigade set up; fertilizer brigade.

1974   Further reorganization of rice-producing brigades, number reduced to 8, with 36 hectares and 124 workers in each. Team for producing reeds and mats set up.

for each kiln, averaged seventeen members. The reforms centralized all the kilns together in one place and brought much of the preparatory work under cover. By 1977 there were 126 workers in the specialized teams detailed in table 4.3. Each team was divided into fixed or permanent groups according to various links in the production process. As in the other branches, there were substantial output and productivity gains. The gross value of production rose from 0.05 m dong to 0.25 m dong between 1970 and 1976. The branch's labor force rose from 52 to 86 over the same period.

In addition to its primary specialization in bricks and tiles, the cooperative had also organized transport brigades. These were apparently very successful and allocated buffalo carts to the fixed rice-producing brigades.[18] During the period from 1971 to 1974, eighteen vehicles with thirty-six workers were permanently let out to brigades. But when the latter were nearly doubled in size after the 1974 reform the system was replaced by temporary groups. A reaping gang plus cart became the basic unit during the harvest. The other subsidiary branch brigades used means of transport belonging to the cooperative (primarily boats and small tractors) to work contracts with the state.

These experiences show the highly detailed division of labor, the diversification, and the specialization that occurred after the introduction of the NMS. It is interesting to note the limits eventually placed upon the integration of production at the cooperative level: pig food was not being grown in the cultivation branch, but by a team in the livestock brigade, while ploughing services were still provided from within the rice brigades.

Vu Thang. The management system in Vu Thang was slightly different from that in Dinh Cong. The machine endowment was even better, and more was made of the strategy of simultaneous development of rice cultivation and the collective pig herd. The basic rice brigades had even less independence, and mechanization had reached the point where 50 percent of land preparation was no longer done by draught animals. Details of the organizational structure were less complete, but the system had apparently evolved as shown in table 4.4.

Table 4.5

Paddy Production and Use, Vu Thang (in *dong*)

	1965	1966	1970	1971	1974	1975	1977
Paddy available (tonnes)	1,075	1,150	1,874	1,976	2,248	1,578	2,065
to livestock	43	52	107	259	386	336	421
of which collective	25	34	73	220	330	283	380
Delivered to state	232	235	517	516	536	450	593
Residual	35	27	73	66	50	61	70
Distributed to cooperators	765	836	1,177	1,155	1,276	731	941
%	71.2	72.7	62.8	58.4	56.8	46.3	45.6
Implied per capita per month kg	17	—	—	—	—	—	21

Source: Nguyen Manh Huan (1979).

Compared with Dinh Cong, two additional brigades provided inputs to the rice brigades: fertilizer and ploughing/harrowing. Responsibility for water supply had not, so far as was known, been allocated to a specific unit. The basic construction brigade would probably have performed some of the services provided by Dinh Cong's irrigation brigade. An additional team produced bricks and tiles.

As in Dinh Cong, this structure was based upon contractual relations with brigades. These were based upon a system of work norms[19] and a categorization of laborers that allowed calculation of remuneration. There were 855 norms and 4 categories of laborer, which ranged from those 10–12 years old, who received 15 kg per month, to category A laborers, with 30 kg per month. The apparent reported workdays per cooperator had risen rapidly, from 180 in 1959–1964 to 346 in 1977.[20]

Apart from revealing the same pattern of diversification and specialization as Dinh Cong, Vu Thang also deserves detailed study because of the particularly interesting way in which it financed pig-rearing. This depended largely upon the enormous paddy surplus to cooperators' subsistence needs (table 4.5).

The data permit calculation of the uses to which the increased output was put. This reveals that between 1965 and 1970 over half the increase had gone to the cooperators and 35 percent to the state. But between 1970 and 1977 the share of the cooperators fell in absolute terms to the benefit of collective livestock and the state. Nothing is known about changes in the composition of remuneration and possible increases in meat consumption. The data also allow some examination of the effects of the bad harvest year of 1975 and the relative priorities of the cooperative's management cadres. In 1975 (compared with 1974) the cooperators were badly hit. Their absolute share of the paddy distributed fell by 43 percent (from 1,276 tonnes to 731) compared with declines of 16 percent in deliveries to the state and 14 percent in supplies to the collective pig herd. The puzzling steadiness in paddy supplies to the household pig herd (56–63 tonnes) may have resulted from strict regulation of that sector or use of low quality paddy.

The subsistence needs of the population were restrained by the slow growth of the commune's population. Throughout the period (1965–1977) the population rose by only 110 (from 3,702 to 3,812), primarily because of emigration to New Economic Zones by 1,490 people. Families that left were given 440 dong and 30 "marks"[21] for setting-up expenses. These people, had they stayed without contributing to output, would have generated subsistence requirements of over 250 tonnes annually—less than the increased deliveries to the state (361 tonnes).

The pigs reared in Vu Thang received a substantial proportion of the increased rice production. Part went to the private sector. Management closely controlled cooperators' pig-rearing, and they had to have permission to slaughter a pig.[22] The larger share went to the collective sties. The cooperative used only a small proportion of its land (2.7 percent in 1977) to grow vegetables as a possibly cheaper alternative to paddy. This implied that much of the manure from the pigs went to the rice fields and should therefore be treated as a direct benefit to the cooperative. On this basis, and with various other "heroic" assumptions, the simplified cost-benefit analysis of table 4.6 results.

Unidentified and so unincluded costs included buying piglets and stud services, vegetable feed, and the costs of food preparation. Labor costs in this branch were usually thought of as being rather high because a greater volume of work and skilled labor was required. This usually increased the need for decent material incentives. In addition, use of private-sector labor was considered to have zero cost to the cooperative, with the rice allocation the only cost to pig-rearing. The financial or workpoint allocation to households for pig-rearing was not known. On these assumptions the exercise implied that the profitability of pig-rearing depended crucially upon the herd size. This agreed with information from other sources, which habitually put the break-even point at about 1,000 head.[23]

The table assumes that the cultivable area was the same throughout the period, and that shadow prices used were correct—for paddy the duty-prices of 0.28 dong/kg before 1966 and 0.32 dong/kg after, and for pork 1.80 and 2.40 dong/kg liveweight: these were the accounting prices cooperatives were meant to use to aggregate nontraded output.[24] The calculation used 10 dong/tonne for animal manure (based upon a figure given in NCKT 91, 73), which gives approximately 2.5 dong/tonne for fertilizer after treatment (my estimate). The volume of fertilizer supplied in 1972 was probably a good proxy for 1970, and similarly 1964 for 1965. The paddy needs of the labor in the branch were somewhat arbitrarily set at 17 kg/head/month, and assumed that the worker had three other people to sup-

Table 4.6

Estimated Costs and Benefits of Pig-Rearing, Vu Thang (in *dong*)

	1965	1970	1977
Benefits—total	12,000	89,000	280,000
Value of meat supplied by cooperative	9,000	66,000	256,000
Value of fertilizer	2,875	22,950	24,395
Identified costs—total	8,000	65,000	170,000
Value of rice supplied to pigs	12,000	34,000	105,000
Rice subsistence needs of labor in branch	3,800	31,000	35,000
Benefits net of identified costs	−6,000	+24,000	+110,000
Size of pig herd (head)	1,193	2,310	3,154
of which collective	83	500	1,484

port. It can be argued that this was not really a cost to the branch, but from the point of view of the hypothetical decision maker it was a real opportunity cost since the labor could be employed elsewhere. In addition, it probably underestimated the real cost of labor in a rich cooperative where incomes were probably well above subsistence.

These two model cooperatives show the changes that should, according to the party, have occurred in cooperatives where the NMS was adopted. They were operational models in the sense that rural cadres were meant to study and learn from them. Management teams from other cooperatives should have adopted their methods to secure unified managerial control of their own cooperatives. Substantial increases in output and mobilized surplus accompanied large supplies of modern inputs and a detailed managerial control over production and distribution. A major element of the shifting local power balances was the erosion of the brigades' economic independence. In production, this corresponded to a general increase in the level of integration of production at the level of the cooperative resulting from an increase in the use of specialized subunits. This developed both outside the rice brigades and, in the form of teams and work groups, inside all brigades. In distribution, it corresponded to the Management Committee's close control over physical output and the social mechanism (the system of workpoints) used to reward collective labor.

But these model cooperatives were not representative of general problems, and the value of the NMS as an overall policy depended upon the more general characteristics of North Vietnamese agriculture and the Agrarian Question it was meant to solve. Further discussion therefore requires some consideration of the theoretical and historical origins of the NMS.

The NMS in Vietnamese Communist Thought

The origins of the complex system of management described above are to be found in the general ideas held by the Vietnamese leadership and the problems that they confronted in agriculture. These will be taken in turn. Consideration of the former was made comparatively simple by the lack of complexity in the ideas used, foremost among which was that of the "Three Revolutions." This corresponded in part to a conventional Marxist idea of the dialectical relationship between the forces and relations of production. The precise origins of these ideas may be found some day by historians, political biographers, and others.[25] Here one can only try to show what they were and their relationship to the North Vietnamese Agrarian Question.

Two absolutely crucial underlying principles were presumption of the historical necessity of the Vietnamese Revolution and certain equally inviolate and unquestioned aspects of it. The former means that the notions frequently appear, by orthodox Western standards, philosophically Idealist. The latter, the meat on the bones of basic ideas, is the fundamental ideological framework. The argument below presents an outline of those elements of the ideological framework that are relevant to cooperativization. It then considers them in detail and the goals cooperativization may be said to have sought to achieve.

The Vietnamese Revolution and the Three Revolutions

The internal logic of the Vietnamese Communist argument demonstrating the necessity of the cooperative form may be expressed as:[26] The National-Democratic Revolution leads the triumph of the "new-type" state lead to The Vietnamese (Socialist) Revolution, which is both "the struggle between capitalism and socialism" and "the Three Revolutions." The latter are "organic parts of the Vietnamese Revolution," i.e., (1) Revolution in production relations; (2) Technical and scientific revolution; and (3) Ideological and cultural revolution. The precise nature of (1) and (2) result from the struggle between capitalism and socialism, i.e., cooperation[27] and state planning in production; these advance in the "forces of production" in the technical revolution. Thus cooperatives are a necessary part of the revolution.

The power of this argument derived from the respect due to its presenters and the values attached to the terms used. Little empirical justification was given for the positions taken, while the holistic nature of the method blurred any possible distinction between means and ends.

A clear statement of the Vietnamese leadership's public position on the nature of development in their country (the basis for the above schematic presentation) can be found in Le Duan's *The Vietnamese Revolution—Fundamental Problems,*

Essential Tasks (1969). It is quite clear that until the late 1970s at least there was little significant change in the general line, while this work shows no fundamental differences from earlier public statements (see, for example, the extracts from Truong Chinh's 1959 speech below). After 1979 it is possible to argue that major alterations did occur, at least in the general atmosphere of debate, and especially in the innate values attached to terms and arguments. Le Duan (1969) expands the above schematic argument. The historical basis for the Vietnamese Revolution was, it is said, the "victorious outcome of the national-democratic revolution and the triumph of a 'new-type' state": "The worker-peasant revolutionary power led by the working class immediately sets about discharging the historical duties of the dictatorship of the proletariat—to carry out a socialist revolution and build socialism" (p. 230).

The greatest characteristic of North Vietnam in this transition period was the "direct advance from agricultural backwardness to socialism without passing through the stage of capitalist development" (p. 229, quoting Ho Chi Minh at the Third Congress of the Vietnam Workers' party, 1960).

A positive role was attributed here to the "new-type" state, for "to build socialism and communism successfully, it is not enough to repress the exploiting classes and other counterrevolutionary forces . . . [this] cannot by itself create the material and technical basis for socialism" (p. 232/4).

The "Worker-Peasant Revolutionary Power" paved the way for the interaction between initial advances in production relations and the development of production. Positive ethical value was attached to socialist production relations[28] in the conception of a dominant struggle between two ways—capitalism and socialism. This dualistic approach reinforced the antagonistic attitude toward the "outside" economy: not being attributable to the "socialist way," it was ethically negative. The slogan that expressed this well was "*Ai thang ai?*": "Who wins, who is defeated?"[29]

This struggle between capitalism and socialism was "a struggle to raise small production to large-scale socialist production . . . to carry out simultaneously [the] Three Revolutions" (p. 234).

The mutual interaction between these "Three Revolutions" was clearly put: "The revolution in production relations, the technical revolution, and the ideological and cultural revolution are the three organic parts of the socialist revolution. They are intertwined, exert influence upon each other, and impel each other forward. The new society, the new man, the new production relations are not the results of any single revolution but the common products of all three" (p. 235).

The idea of the revolution in production relations was comparatively broad, involving, apart from "instituting the collective socialist ownership of the means of production," collective mastery over economic management and distribution of the products of labor. The technical revolution was also comparatively wide-ranging and included the idea of piecemeal advance without new technology.[30]

The two were closely linked:

> The technical revolution is closely bound to the revolution in production rela-
> tions, and the two exert reciprocal influences in a dialectical way. The latter
> paves the way for the former and creates socioeconomic premises for pushing it
> forward; conversely, the former consolidates the fruits of the latter and creates
> material-technical premises for ceaselessly perfecting socialist production rela-
> tions. (p. 238)

An apparent corollary of this was that the collective character of ownership of
the means of production could not be expected to be stable if production did not
develop. A major contradiction could then arise between the "advanced" production relations and the "backward" level of development of production. In
1974 Le Duan recognized this in "The New Stage of Our Revolution and the
Tasks of the Trade Unions": "a cooperativized agriculture can only be viable if
based upon large-scale industrial production" (p. 350).

The basic rationale given for agricultural cooperatives was therefore that they
were a necessary part of a necessary historical phenomenon. Other justifications
have certainly been presented,[31] and in his keynote speech on cooperativization
in 1959, then First-Secretary Truong Chinh summarized the arguments as follows:

1. The need to prevent the "capitalist road" winning in agriculture.
2. The need to abolish exploitation and improve the living conditions of the
masses.
3. The need to reinforce the worker-peasant alliance which was the basis of
the state.
4. The need to construct socialism in the North as a whole. Here he emphasized the problem of private trade and the relationship between traders and rich
peasants.
5. The need to overcome the deepening contradiction between state industry
and an unplanned peasant agriculture.
6. The need to ensure supplies to the state for export and for distribution.
7. The need to secure the North and ensure the success of the struggle to
reunify the country.[32]

Here it is possible to observe clearly the power of the "blanket solution,"
which allowed the use of a mixture of political and economic arguments. The
holistic nature of such thinking was closely related to belief in the historical
necessity of the revolution. Truong Chinh expressly pointed out that the party had
always maintained that the socialist revolution followed the national-democratic
revolution (p. 23). This implied that the above list of factors only revealed
different aspects of an inevitable process.

More concrete problems in the economic sphere were dealt with in the context of the worker-peasant alliance (no. 3 above). Thus Le Duan's statement: "[In the present socialist revolution] the foundation of the worker-peasant alliance [is] agricultural cooperation and socialist industrialization."[33]

In this alliance the workers took the lead under the guidance of the party bearing their name. The workers, identified with industry—and especially state industry—supplied the peasants with means of production and consumer goods and played a leading role: "their" party led the peasants to cooperativization. The peasants supplied industry with prerequisites to the construction of socialism: "the labor force, consumer goods, primitive accumulation, and the market."

These are very broad conceptions. They possessed considerable power by virtue of their ethical content. The moral commitment to participation in the revolution corresponded to the fulfillment of certain duties and responsibilities; eloquent testimony to this came from a Vietnamese economist who was clearly not a party theoretician:

> The cooperative is a collective socialist economic organization freely set up by the peasantry under the assistance of the state. *Thus in each situation the state must help the cooperative.*
>
> The state is the property of the proletariat, the workers. The peasants are the friends and allies of the workers, *so the proletariat must help the peasantry.*[34]

This shows the two main characteristics of fundamental Vietnamese Communist thinking about the role of cooperatives in the Vietnamese Revolution: its reliance upon historical necessity and its inherent ethical value. It is important to note that the cooperative movement was not seen as a means to some explicit set of social ends, a means that had no value in itself. This meant that indigenous critical analysis inevitably found itself limited in scope, since criticism should have been posed in terms of deviation from the party "ideal" rather than the attainment of accepted goals (e.g., output maximization subject to distributional constraints).

Such thinking is not perhaps surprising for people leading a National Liberation struggle, first against French colonialism and then against U.S. military might. Once identified as part of the Vietnamese Socialist Revolution, there must have been strong arguments that the cooperatives had to be supported. But in practice and in peacetime such idealistic thinking can run into a number of difficulties. In wartime, the stress on ethics and necessity, as elsewhere, helps mobilize energies for sacrifice and determination required by the common struggle. Dreams are made of this. But there are many sorts of idealism, and it is not certain that these grounds truly justify the NMS. Reality must start to butt in when people want to attain simple but concrete goals. This may happen in peacetime situations when a means-ends dichotomy is pushed forward by people who want to get richer—or simply no longer rely upon food aid to feed their children. They then see less reason to support unpopular policies that were clearly nonimplementable. In North Vietnam such idealistic ways of thinking had received a strong boost from wartime "hypernationalism," and this perhaps partly explains why they were continued with for so long. Criticism of policy could easily be attacked for being unpatriotic. It was not until the economic crisis of the late 1970s and the 6th Plenum of August 1979 that characteristic Vietnamese pragmatism started to surface in open public debate.

The Cooperative Movement in the DRV

The origins of the NMS must be sought in the particular problems presented by the operation of cooperativized agriculture in North Vietnam. Here certain points are of importance. First, the NMS was a solution to problems presented by the collectivized system as it was operating in the late 1960s and early 1970s; it was not a solution to the problems of the uncollectivized system. Second, the strongly idealistic nature of the basic ideological framework meant that policy nonimplementation itself was a vital issue. If cooperatives were a historical necessity and inherently good then they had to be made to work. If they continually and spontaneously became nominalized, this cast considerable and intolerable doubts upon their historical necessity and innate moral worth. Because, as is argued below, the great mass of cooperatives were indeed nominalized, the Agrarian Question presents itself as the deep contradiction between everyday reality and certain basic tenets of Vietnamese Communism. Although these tenets remained generally valid—or unchallenged—the overriding stress had to be upon implementation of basic policy. The tenor of discussions had always to be the way in which cooperatives could be made to function properly. To ask whether they were playing a useful part in society (and if so, what?) was to question the unquestionable: by definition they belonged to the Vietnamese Revolution.

From this perspective, undue stress on developmentalist issues somewhat misses the point. There are many ways of raising output, improving distributional equity, increasing levels of mobilized surplus, and so on. Modified versions of the NMS were quite capable of doing so. But even if they did, they were unacceptable. Evaluation of cooperatives had in principle simply to compare their operation with the models described above. In practice, this led to revealing and very human dilemmas. Some authors sought to escape the official mould, reintroducing developmentalist issues and finding it hard to condemn unorthodox managements that nevertheless raised economic activity within the collectivity, apparently to general benefit.

A further point is that policy nonimplementation had been widespread from very early on. In the context of the DRV's overall economic management system, this had provoked increasing debate toward the end of the First Five-Year Plan. Systemic policy nonimplementability created particular problems for North Vietnamese legalists. North Vietnamese policy-makers, it has been argued else-where, tended to balance the extreme idealism that accompanied wartime hyper-nationalism with an acceptance of widespread policy failure that probably rein-forced their idealism.[35]

Early Problems. The existence of problems was recognized from the very earliest days of the cooperativization movement. Two separate campaigns at-tempted to deal with these:[36]

1. Stages 1 and 2 of the campaign to "improve cooperative management, improve methods of production, and develop agriculture thoroughly, strongly, and firmly." These started in June 1963 and October 1965 (end-dates unknown).

2. The campaign "to reorganize production from the base in the direction of socialist production" from August 1974, which used the NMS. It was preceded by a "period of experimentation" (1971–74) with methods of "leading agricul-ture toward large-scale socialist production."

The campaign's first stage coincided with a broad shift in policy emphasis during the failed First Five-Year Plan (1961–65). Macro-tensions resulting from the overly ambitious and forced development program were pushing up food prices as agricultural output growth lagged behind demand from the increased nonagricultural labor force. By 1963 at least, many cooperators wanted to leave their cooperatives.[37]

These early drives to improve cooperative management were based upon the evidence from such studies as Bui Huy Dap's monumental work *Rice in North Vietnam.*[38] This purported to show the existence of considerable scope for im-provements in yields and output without substantial increases in supplies of modern inputs. Bui Huy Dap pointed out the overriding need to improve land yields that resulted from the extremely limited capacity for extending the cultivat-ed area. This could come from increases in both the sown area and the yield per harvest; potential existed for raising the number of harvests, improving water control, and so on. He also emphasized the need to concentrate, within the labor process, on those tasks that acted directly on yields. The piecemeal nature of possible gains helped explain the emphasis put upon the wide-ranging content of the technical revolution in the more abstract formulations discussed above. But stress on the scope for yield increases without additional industrial inputs was still consistent with the neo-Stalinist priority on investment in modern industry.[39]

The first campaign appears to have had two main aims: an extension of the area of operation of the collective economy, and an emphasis upon managing relations between the cooperative and its subunits through use of the III-point contract system. Suitably modified, both of these are to be found in the NMS. Extension of the collective economy had two elements. First, the smaller cooperatives were steadily amalgamated to form cooperatives approaching village or commune level. Second, the lower-level cooperatives were transformed into higher-level cooperatives, a process that was approximately concluded by 1968.[40] This oc-curred despite the growing feeling among many close to policy makers that the major contradiction in the countryside was that between the "weak" but nevertheless still socialist production relations and an agricultural technology that had not really changed since land reform. Methods of production, it was said, still essentially corresponded to those of petty producers. In retrospect, this could be justified by the same basic argument as that for cooperativization per se: "reinforcing socialist collective property and increasing the size of cooperatives are two aspects of the same problem, with the effect of simultaneously reinforcing and purifying collective property, and creating the conditions for a strong development of the collective economy" (Dinh Thu Cuc, p. 40).

The attitude toward petty production was highly antipathetic, and many statements by top leaders could be found attributing difficulties with the cooperatives in particular and the entire neo-Stalinist model in general to the persistence of "small-scale production" (san xuat nho). This plays a major role in helping to explain attitudes toward the private sector and the cooperators' own-account activities. The shift to higher-level cooperatives in principle constrained development of the "minor household economy." The importance of this hostility in the formulation of the NMS is dealt with below.

The III-point contract system provided a first basis for relations between the subunits of the enlarged higher-level cooperatives and the cooperative's Management Committee by stipulating the levels of labor and nonlabor inputs as well as output. Its importance was frequently emphasized by General Nguyen Chi Thanh, head of the party's Central Agricultural Committee for part of the 1960s.[41]

Deep problems surfaced in the late 1960s, when U.S. bombings and severe macroeconomic problems were producing the shift to food import dependency that was evident by the end of the decade. Despite the rapid growth in both the size of cooperatives and the influence (at least in principle) of the collective sector, cooperative managements were not using prescribed methods. The known potential for relatively costless technical improvement remained unutilized. Everyday experience pointed to considerable absolute waste. "Incompetent" management teams were not using effectively the greater powers they had been given. A historian writing in the late 1970s and surveying the results of cooperativization could authoritatively write: "Looking back over nearly twenty years [it is clear that] . . . in many cooperatives . . . a number of cadres lack the collective spirit and embezzle and waste [collective property]" (Dinh Thu Cuc, p. 48). And, "The situation of cooperators taking over collective land without permission is . . . generalized and has been so for many years" (p. 40).

In July 1969 the Central (party) Agricultural Committee report on "10 Years of the Movement to Cooperativize Agriculture" revealed, or so it seemed, the predominance of the collective property-form in agriculture: 80 percent of cooperatives were higher-level, and these contained 92.4 percent of cooperator family-units and 91.7 percent of the cultivated cooperative area.[42] The numbers of hamlet-level cooperatives were being reduced, and there were 550 cooperatives

(2.5 percent of the total) at the commune level. Yet the collectively farmed area was actually thought to be declining in some areas: "Both of the waves to improve cooperative management [had] raised the problem of how to relocate land given out incorrectly to cooperators. . . . This problem was an extremely important phenomenon in almost all cooperatives" (Dinh Thu Cuc, p. 40). This was especially notable in Nghe An, which had been badly hit by U.S. bombing, but even in the comparatively peaceful Thai Binh the area cultivated collectively fell by 12.6 percent in fourteen years (1960–1973), and by 24.8 percent on a per capita basis. Similar problems were quoted for Hai Phong and Bac Thai.[43] But not only land was under pressure: "care of draught animals, implements and tools, means of production, and the material-technical base of the cooperative were still not properly looked after, which influenced the stability of the cooperative" (Dinh Thu Cuc, p. 41).

This source, while speaking "from the shoulder" and in an extremely authoritative journal, only says more loudly and emphatically what had probably been obvious from the earliest years of the 1960s. Although wartime popular mobilization had been successful, the party line was not fully implementable in the rural areas. This confirms the point made above, that acceptance of policy nonimplementability was an inherent part of Vietnamese Communist orthodoxy, probably reinforcing its characteristic idealism.

Attempts to rectify these problems aimed at improving the quality of management. In July 1961 the 5th Plenum had affirmed that "Reforming and improving the management of cooperatives is the most important element in developing production" (Dinh Thu Cuc, p. 48). This opinion does not seem to have altered in later years. A sine qua non of effective management was a minimum level of long-term planning, but managements tended only to plan "for the next harvest" (p. 43). There had been some limited success with attempts at production diversification during the earlier campaign, but by the late 1960s, 53 percent of cooperatives still lacked collective pig herds, 43 percent were not rearing fish, and 45 percent had still not yet diversified into industrial crops and fruits. None of these advances required significant supplies of outside inputs, but they required some degree of planning, which "was still progressing slowly and incompletely" (p. 40).

Despite all these difficulties the average size of cooperatives had continued to increase. Over 70 percent were at commune level by the mid–1970s.[43] But this exacerbated the problem by effectively decentralizing cooperative management:

> Once the cooperative has been expanded and has made the advance to the higher-level, it has the possibility of developing the forces of production, *but this does not happen frequently*. One important reason for this is the "dispersed" character of the management of production . . . for a long period (*more than ten years*) the method of management has been to allocate out to the brigades . . . [who] have held land, labor and [production] expenses and have organized production

by themselves, frequently independently of the cooperative's plan. (Dinh Thu Cuc, pp. 44–45; emphasis added)

Remuneration to labor was intended to follow the Leninist principle of "work according to capacity, pay according to labor." In principle, this was meant to allow a harmonious balance between individual interests and those of society. The 5th Plenum mentioned above established that in any cooperative (high or low level) the income distributed according to labor was to exceed 50 percent of the total (Dinh Thu Cuc, p. 48). This required a minimal system of work norms, yet "For over ten years now, apart from a number of cooperatives, almost all produce without regulations and without work norms" (p. 49).

The implications for the nature of the cooperative movement of this and other evidence are immense. In terms of the typology introduced in chapter 3 it implies that the great majority of cooperatives were of the nominal type. It follows that cooperative reform was primarily a deliberate attempt to change the attitudes and activities of management teams. The implied existence of a vast gap between the prescriptions of the party and the practice of the rural periphery[45] was consistent with the traditional functions of local administrative structures as "protective intermediaries" between local interests and higher levels.

Some indication can be had of the likely numbers of model and the more progressive active cooperatives, although the evidence is far from conclusive. First, there are the official data on the size of the DRV collective pig herd. As has been seen, this particular method of production diversification was a key part of the NMS. As even the model cooperatives showed, its profitability was in practice somewhat problematic. The average collective herd in the sample of cooperatives considered below was around 650. Although these were all rather advanced cooperatives whose managements were comparatively active, they by no means displayed the model cooperatives' full and proper respect for the norms of the NMS. Taken with the DRV total of 640,000 over the two years 1974–75,[46] this implied that there were little more than 1,000 advanced cooperatives out of a total of 17,000 at that time, or just over 5 percent.

Second, there was the official Vietnamese classification of cooperatives. Frequently encountered in documentary sources, this divided cooperatives into three groups—good, mediocre, and bad. No basis for this was ever given, and it is not even known whether the basis was stable over time. The proportion in the "good" category rarely exceeded one-third.

A third basis for categorizing cooperatives appeared in the late 1970s, where "10.8 percent of the total number of cooperatives" were said to be "like Vu Thang."[47]

Although none of these arguments is particularly convincing, the lack of any concrete evidence to the contrary can only lead to the strong possibility that "nominal"-type cooperatives were around 75 percent of the total in the mid-1970s. While this ratio is obviously not accurate to within plus or minus 10 percent it is reasonably certain that nominal cooperatives were in the great majority. Nominalization was therefore widespread, and probably the norm.

The Agrarian Question in the Early 1970s. If this characterization of Vietnamese agriculture is correct then the full nature of the problems facing policy makers can now be fully appreciated. The collectivized system was providing *bo doi* for the army and high levels of rice procurement. But it was consistently flouting basic tenets of Vietnamese communism. Any realistic assessment of socioeconomic activity in rural areas had to conclude from cooperators' everyday behavior that the party's orthodox notion of cooperatives was neither historically necessary nor innately good. If it had been, peasants would hardly have spent so much energy trying to avoid implementing central directives. Cooperative managements would have found it easy to get cooperators to work for the collective, and there would not have been the continual—and frequently successful—nibbling at collective assets such as land.

The dilemma faced is clear. Despite the obvious inadequacies and disobedience of local cadres, getting rid of them posed immense problems. Even if the central leadership did contemplate such a solution, they could hardly be purged during wartime. In any case such a venture would have been extremely unpopular and risky. The central leadership could hardly set about replacing the cadres of 75 percent of the cooperatives by direct administrative means. To some extent, therefore, the NMS can usefully be seen as a way of producing such a purge indirectly, by creating a power base within cooperatives for a new technocratic group.

Part of the problem inevitably resulted from the overlapping of political and economic functions in rural administration during a period of macroeconomic disequilibrium and great demographic pressure upon subsistence. Chapter 2 showed how the cooperatives were required to act simultaneously as both political and economic units. In the economic sphere they functioned as both mobilizers of surplus within the national economy and an important source of cooperator income. In the political sphere, however, they were a major source of any authority exercised by the lowest levels of the party and rural administration. This supported efforts to maintain morale and troop mobilizations. But at the same time limits were placed upon the extent to which the wishes of the cooperators could be ignored: of crucial importance when the thriving free market and pressure upon subsistence combined with the shortages of state-supplied goods to make collective labor relatively unattractive.

Policy Measures before the Thai Binh Conference. The obvious difficulties with the operation of the economic management system of the DRV prompted debate and various attempts at reform from the early 1960s. From the late 1960s these pressures seem to have grown, until by the very early 1970s strong attempts were being made to do something about the situation. In agriculture this had started with the push to improve cooperative managements introduced from 1963 in the face of rising popular discontent. By the early 1970s the operation of the

Stalinist system of administrative goods supply was subject to clear criticism, reportedly at three key plena of the period—the Nineteenth in February 1971, the Twentieth in April 1972, and the Twenty-second in April 1974.[48]

The Nineteenth Plenum concentrated upon agricultural problems, advocating reinforcement of local administrative and party organizations. It also began to push strongly for the diversification of cooperative activities into livestock and industry.[49] The plenum also approved a target of 30 percent nonrice staples production.[50] A year later the Twentieth Plenum looked for a reinforcement of party control over the economy, looking for "the abolition of administrative management of supply and implementation of management according to socialist business methods; getting rid of management according to dispersed artisanal and small-scale production methods; constructing the management methods of large-scale industry in order to stimulate the process whereby the national economy will advance from small-scale production to large-scale socialist production."[51] By this stage efforts were clearly focused upon economic management methods consistent with the basic themes of control and centralization that had long been advocated by top leaders. Thus the Twenty-second Plenum, held over a year after the signing of the Paris Agreements, reasserted the party's hostility to small-scale production[52] and confirmed the commitment to the underlying drive to control that was such a key element of the NMS.

At the level of the macro-economy the price problem remained an obstacle to controlling resources in the socialist sectors. A decree of early 1974 announced the abolition of the free market in staples;[53] the government also announced changes in prices and subsidies throughout 1974 that sought to reduce the gap between free market and state prices.[54] The evidence from the grass-roots cited below suggests that these measures had little effect on the overall adverse balance of incentives facing Management Teams when trying to induce cooperators to work for the cooperatives.

The Thai Binh Conference. The Thai Binh Conference of August 1974 marked the end of a period of "experimentation with management methods." Inauguration of the campaign "to reorganize production from the base in the direction of socialist production" established official commitment to the NMS. The extent of the difficulties then being officially admitted can be seen from the speech by Hoang Anh. His criticisms of the cooperative movement confirm the various problems already mentioned. The cultivated area had "seriously fallen," while yields and livestock numbers had "stagnated." His major attack on the state of "production relations" was upon the shortcomings of management, especially the "loose" attitude to collective property, the "irrational" and "piecemeal" approach to the organization of production, and the inequitable pattern of distribution. In particular the widespread tendency toward nominal cooperatives was implicitly revealed, for "the majority of cooperatives . . . while seeming to be production units of several hundred hectares, operate in reality in production groups [that] organize for themselves the utilization of land, labor tools, and materials, and the distribution of the product. . . . In the mediocre cooperatives . . . the cooperative avoids performing certain tasks and gives them . . . to the families of cooperators."[55]

Almost identical points occur in the key piece by Le Duan in his *Toward a Large-Scale Socialist Agriculture* (1975), which emphasized the need for cooperatives to "take in hand the means of production" and "discard the system of organization of production and distribution of profits by individual production brigades . . . by unifying the utilization and management of land on the scale of the whole cooperative."[56]

Almost all the characteristics of nominal cooperatives were apparent from these criticisms: poor utilization of technical improvements and other ways of increasing yields in a supply-constrained economy, decentralized management, extensive privatization of collective assets (especially land), and indications of nonobservance of "socialist norms" by local cadres ("democratic principles . . . are not respected").

The South and the 1976 Reaffirmation of the NMS. In 1976 the Twenty-fourth Plenum analyzed the situation in recently liberated South Vietnam and criticized its economic development for being dependent on foreign imports and still predominantly based upon small-scale production.[57] These, of course, were criticisms that were especially valid for the state of the Northern economy at that time. The top party leadership reaffirmed its support for the thinking behind the NMS and the Government Council issued decree 61-CP in April 1976, "Toward Large-Scale Socialist Agricultural Production."[58] This reaffirmation of the appropriateness of the NMS committed the country to it for the coming period, during which there steadily developed a major economic crisis that was given a final stimulus by the aid cutbacks of 1978–79. Out of this process—with some temporary setbacks—came the so-called output contracting system formalized in 1980–81, which effectively decentralized prescribed cooperative management structures. This is described at greater length in part IV, but from 1974 until 1979 the NMS remained official policy.

The NMS was essentially a response to the widespread prevalence of nominal-type cooperatives, and its emphasis upon managerial reform was quite understandable so long as the party center retained the initiative and a commitment to its own idea of "socialist advance," within which a continual reinforcement of socialist relations of production was a sine qua non.

Implementation of the NMS—Likely Consequences

Managerial reform along the lines of the NMS can now be understood in terms of the typology of cooperatives presented in chapter 3. In particular, the NMS's key "three principles of management" may now be reinterpreted:

—Unification of economic activity: the Management Committee, not the brigade leadership, must control collective production and distribution.

—Unification of management: land, labor, and means of production must be controlled by the committee, not by the brigades, and certainly not by the households.

—Unification of distribution: collective output and imported consumer goods must be allocated by the Management Committee, and not siphoned off by any other agent.

Management "reform" would start from a structure close to that of the nominal cooperative. It would often probably result from or start with the amalgamation of a number of smaller cooperatives. Reform would introduce conflicts and tensions as the pattern of dominance came under pressure. A successful reform generated one or other of the "advanced" types (i.e., model or active). An unsuccessful reform left the Management Committee as powerless and irrelevant as it had been before. When observing the different outcomes of these exercises we can examine both the factors determining their success or failure and the development of production and surplus mobilization. Thus direct material interests can be seen in operation.

Patterns of Conflict

An attempt to implement the NMS in a nominal cooperative would be characterized by potential conflicts corresponding to the differences between these and advanced cooperatives:

Conflicts over means of production. Of these, land, implements, fertilizer, seeds, and livestock would be of great importance. This would have consequences for the household sector's independence when privatization of cooperative assets had been extensive and when regulation of own-account activity was being extended by a newly active Management Committee.

Conflicts over output distribution. Reformers would seek to replace brigade-level distribution by cooperative-level distribution, and rich brigades would tend to suffer as resources were centralized. Expropriation would be attacked.

Conflicts over labor and control of the labor process. Reformers would attempt to ensure effective labor mobilization and good labor discipline.

Such attempts to replace local cadres by "progressive"—technocratic—elements could lead to conflicts. Older cadres entrenched in the production brigades could ally with their cooperators in a struggle for control over land, labor, and means of production. Such a struggle was not conducive to the social unity required for the war effort. Nor were the older cadres likely to be entirely without support in higher levels. The dilemma faced was of the boot-strap variety: attempts at reform had generally to use the existing structures which were themselves the object of reform. From this followed an emphasis upon outside cadre teams. Overcoming the resistance of local cadres to the use of prescribed methods of economic management in the conditions of the 1970s proved, in the event, largely impossible. The NMS was nevertheless an attempt to do so.

If the reforms began to succeed, the forum for conflicts would become much more dynamic, with continuing change in the system of production. The creation of functionally and branch specialized brigades would herald the introduction of growth processes based upon diversification and increased surplus generation. This would itself reinforce the position of the new leadership by a general shift in the basis of dominance from distribution to production. Conflict would often be posed in terms of old versus young (the new group would be made up of younger and better educated cadres) and insiders versus outsiders (the new group would frequently be dominated or influenced by a team of outside cadres as part of attempts to bypass the existing rural administration).

The factors influencing the success or failure of the exercise were summarized at the end of chapter 3. Such changes could occur spontaneously, but two crucial exogenous factors were, first, the presence or otherwise of direct outside influence, and the precise form it took; second, the favorable or unfavorable allocation of inputs from the state, including the possibility of reduced procurement.

Economic Consequences

The economic consequences of a successful managerial reform (understood as a transition from nominal to advanced status) would be:

1. Strict regulation of the household sector: reduced sales to the free market and improved supplies to the collective (labor, means of production, etc.).

2. Improvements in yields resulting from improvements in techniques, the use of more inputs from inside the cooperative, and direct accumulation.

3. Increased deliveries to the state, resulting from both a stricter control over output and diversification and increases in production.

Conclusions

This chapter has used the framework established earlier to examine the origins and content of official policy toward the agricultural producer cooperatives. The NMS was shown to be dominated by a desire to unify control of economic activity within the cooperative in the hands of the Management Committee.

In its practical operation this corresponded to a sustained attack upon the rice monocultural brigades typical of nominal cooperatives. Essential inputs to rice cultivation were now obtained from specialized brigades or teams, and a rapid diversification of production occurred. The apparent success of the NMS in model cooperatives was not, however, officially attributed to the highly favorable resource endowments but to the management system. In principle the NMS did not require such additional outside inputs and could lead to a "self-reliant" approach: the emphasis upon pig-rearing was a part of this.

The dominant characteristics of official thinking were the historical necessity and the inherent ethical value of the cooperative movement as part of the Vietnamese Revolution. Confronted with the evident and widespread nonobservance of statutory norms and an erosion of collective authority, the response to the generalized phenomenon of nominal cooperatives was the renewed attempt to strengthen the cooperative. In ideological terms, this policy simply reasserted the historical necessity and inherent value of the cooperative.

Starting with the transition to higher-order cooperatives and the early use of III-point contracts, agrarian policy culminated in the detail and sophistication of the NMS, with its ability to curb brigade independence in both production and distribution. This process led to increases in the size of cooperatives through amalgamation. It also continually gave greater emphasis to the collective sector rather than to the household. It arguably helped to create the conditions for a further divorce between principle and practice by setting up comparatively powerless Management Committees above the older and well-established autarkic rice-monocultural brigades. Reform had then to cope with the resulting nominal structures in a situation where widespread shortages of goods within the state-controlled distribution system combined with a highly profitable free market to make few cooperators eager at the margin to work for the collective. The resulting conflicts form the basis for much of the discussion in subsequent chapters.

Part III

Life at the Grass Roots

5

The Sample in
National Context

Policy implementation at the micro level—the amalgamation of cooperatives and the creation of new managerial structures—generated a need for knowledge of experience in other cooperatives. This in turn produced a variety of studies and reports which form the basis for the materials used here. They were written for domestic consumption, and the extent of the difficulties revealed allows more confidence to be placed in them than in the usual reports from foreign visitors to "Potemkin" cooperatives. But their focus was usually the problems encountered in the so-called advanced (*tien tien**) cooperatives, and although in the present state of research they are of great value, their scope is inevitably limited. The two sources of material are a series of unpublished reports written by economics students at Hanoi University[1] and various articles in the journal *Nghien cuu kinh te* (Economic Research).

To overcome difficulties of incomplete coverage, the sample chosen concentrated upon those cooperatives written about from a number of different viewpoints. To take one example, the cooperative Van Tien in Ha Nam Ninh province was covered by three pieces. These totaled nearly 40,000 words and dealt in turn with problems of labor allocation, the cultivation branch, and the development of the subsidiary branch.

The material's objectivity varied. It was possible to observe authors whose attitude to deviations from the prescribed ideal[2] was far less unreservedly critical than others, especially when the results of the methods used in terms of production and surplus mobilization were clearly favorable. The explanations given for failure typically varied from the "lack of unity of the cooperative's leadership" to clear indications of opposition from the cooperators (in one case heightened by their being predominantly Catholic).[3] The material taken from *Economic Research* had been approved for a wider audience, and criticism was both clearer and sharper, though more also usually put in terms of deviation from the prescribed ideal. My discussions in Hanoi during 1978–79 were a valuable supplementary source. The bibliography lists all the sources used.

The Cooperatives of the Sample

The reports dealt with a number of the so-called advanced or leading cooperatives in North Vietnam in the early and mid-1970s, with some historical reference. The language and basic style of presentation corresponded to that of the Statutes. All were delta cooperatives with no minority population that were predominantly engaged in rice cultivation. All were based upon preexisting administrative units, either at commune or, in the case of one cooperative, the supracommune level. They were without exception the result of the amalgamation of smaller cooperatives from those same administrative units.

As expected, working members of the cooperatives belonged to a variety of organizational subunits. The brigade was the most usual, but teams and groups also existed both as levels equivalent in some way to brigades and as brigade subunits. The brigades were usually simple rice producers, but this type was often accompanied by others. The prescriptions of the Statute were followed in that brigades tended to have a certain allocation of resources in the form of land, labor, and means of production. Their members also received workpoints, which apparently entitled them to a share of collective product and other consumer goods. But the precise nature of relations among cooperators, brigades, and higher levels within the cooperative was extremely variable. This corresponded to the distinction between nominal, active, and model cooperatives. Such differences had great influence upon the development of production. This typically involved an initial investment in water-control works, followed by the use of better seeds and increased fertilizer inputs and the use of sidelines such as livestock, vegetables, and artisanal production. Sidelines could rapidly increase incomes although shortages of raw materials were a severe constraint upon artisanal production. Mobilization and control of collective labor was a crucial problem; cooperators retained small plots of varying extent for their own use.

The cooperatives studied were far from uniform. At one extreme was the cooperative My Tho (Binh Luc district, Nam Ha [sic]), which, before reform, was hardly more than a front used by the local leadership for their own ends. At another extreme were the two model cooperatives Dinh Cong (Thanh Hoa) and Vu Thang (Thai Binh), both of which were well-publicized showpieces of the party and possessed highly developed, centralized management systems as dem-

Table 5.1

Population, Labor, and Other Characteristics of the Sample Cooperatives

	Model				Active				Nominal
Cooperative	Dinh Cong	Vu Thang	Dai Thang (Xuan Thuy)	Dai Thang (Quynh Luu)	Van Tien	Hai Nam	Giao Tan	Thang Long	My Tho
Population	3,800	3,800	8,200	2,900	5,200	5,700	5,000	7,000	2,900
Families	800	950	1,900	650	n/a	n/a	n/a	n/a	n/a
Labor force[a]	1,250	1,400	3,150	1,450	2,350	2,350	1,750	1,600	ca. 1,050
% of population	33	37	38	50	45	41	35	23	ca. 36
Cultivable area (ha)	396	287	613	311	212	452	375	ca. 320	413
ha/head[b]	0.10	0.07	0.07	0.11	0.04	0.08	0.07	0.05	0.14
Rice land (ha)	245	250	451	300(?)	183	325	205(?)	n/a	315
Rice yield (t/ha)	8	7.5	4.5	6	7	6	6.5	5.5[b]	3
Collective pig herd	1,200	1,500	250	n/a	350	500	120	n/a	nil
Private pig herd	n/a	1,650	1,800	n/a	1,960	1,700	n/a	n/a	n/a
Estimated paddy production per capita per month (kg)	43	41	20	52	20	28	22	20[b]	27

[a]Numbers in principle subject to labor duties, and not equal to the number actually working for the collective—except Thang Long cooperative—"workers active in the cooperative."
[b]Hectares of cultivable land per head of population.
(?)—Source not clear.

onstrated above. In addition to these three were six others in intermediate situations:

Van Tien (Hai Van commune, Hai Hau district, Ha Nam Ninh)—a very densely populated cooperative that had developed artisanal production and cultivation in an idiosyncratic manner based upon extremely efficient nonmechanized rice cultivation. Nearly 80 percent of the population was Catholic.

Dai Thang (Xuan Thuy district, Ha Nam Ninh)—a cooperative suffering from the effects of weak management exacerbated by the close proximity of a large market. Sixty-five percent of the population was Catholic. Development had benefited from the tradition of jute cultivation in an old French plantation area.

Dai Thang (Quynh Luu district, Nghe Tinh)—a cooperative close to the model type in the heartland of the revolution, with good access to communications and great potential for the development of subsidiary lines. A focus for the standard process of reform, with outside cadres and a very strict system of penalties and bonuses.

Hai Nam (n/a, but probably in Ha Nam Ninh)—a cooperative relatively well-endowed with machines that had seen some development of production. But the management system was relatively slack and artisanal work was based upon the brigades, which remained comparatively independent.

Giao Tan (Xuan Thuy district, Ha Nam Ninh)—on the sea, a typical cooperative of the intermediate active type, with diversification and development occurring slowly with relatively independent brigades and decentralized artisanal production. Authority at the Management Committee level permitted a system of distribution to "priority families" throughout the cooperative.

Thang Long (Kinh Mon district, Hai Hung)—a mechanized cooperative where sales of various vegetables and other commodities had reached very high levels. But the degree of labor specialization had not gone beyond "simple cooperation appropriate to the level of technology."

Basic data on these cooperatives are set out in table 5.1. The cooperatives of the sample were all relatively large, with above-average yields and population densities. They often had well-developed pig herds and had made some progress in diversifying away from rice monoculture. But labor mobilization levels were absolutely low: typically 75 percent of the registered labor force and 50 percent of the available labor-days.

Table 5.2 shows the geographical situation of each cooperative and the history of amalgamations that had produced it. The large size of these cooperatives (around five times the national average[4]) is probably largely to do with the role played by cooperative amalgamation. Growth was concentrated during two periods (around 1967 or 1973), and from two distinct levels. This periodization seems broadly to correspond to different stages in the evolution of policy toward the cooperatives (see chapter 4). It is possible that the cooperatives of the sample had seen some efforts at reform during the later stages of experimentation that preceded the Thai Binh Conference of August 1974. The existence of two distinct

Table 5.2

Cooperatives of the Sample—Geographical Situation, Size, and History

Cooperative	Status	Province	Delta Population	Population[a] per 1,000 sq. mile	Corresponding administrative unit	Number of constituent subunits	
						A. At	B. Oldest
Dinh Cong	model	Thanh Hoa	Ma River 3,800	1.0	Commune(?)	2 (1969)	11 (1959–64)
Vu Thang	model	Thai Binh	Red River 3,800	1.3	Commune(?)	4 (?-pre-1965)	14 (1959–61)
Dai Thang (Xuan Thuy)	active	Ha Ham Ninh	Red River 8,200	1.3	Commune	3 (1973)	n/a
Dai Thang (Quynh Luu)	active	Nghe Tinh	Ca River(?) 2,900	0.9	Commune	7 (1967)	n/a
Van Tien	active	Ha Nam Ninh	Red River 5,200	2.4	Commune	11 (1967)	16 (1960)
Hai Nam	active	n/a	n/a 5,700	1.3	Commune	3 (1973)	n/a
Giao Tan	active	Ha Nam Ninh	Red River 5,000	1.3	4 villages[b]	2 (1974)	4 (1969)
Thang Long	active	Hai Hung	Red River 7,000	2.2	Commune	2 (1966)	6
My Tho	nominal	Ha Nam Ninh	Red River 2,900	0.7	Commune	3 (1973)	n/a

[a]Population per 1,000 sq. mile of cultivable land (as in table 5.1).
[b]The cooperative consisted of four villages, with the three from a single commune joined with one from another commune.
(?)—Source not clear.

levels from which amalgamation could occur probably reflects the underlying administrative-geographical distinction between the village and the subvillage hamlet. An amalgamation of two to four cooperatives to commune level was probably an amalgamation from village level. An amalgamation from a number much more than that is more likely to be from hamlet level. Note that hamlets usually provided the population basis for the autarkic rice-producing brigades of an "unreformed" cooperative.

Comparisons with National Averages

Table 5.3 compares the main characteristics of the sample with national averages. Although data gaps often necessitated the omission of one or two of the sample cooperatives from the calculation this does not greatly affect the conclusions of the argument. This should become clearer in later sections. Table 5.3 shows clearly that these cooperatives were not only far larger than the average, but also comparatively densely populated; this had to some extent offset their generally better land yields. The relatively large average size of their collective pig herds illustrates their relative success in implementing this element of the NMS.

Data omissions plus inadequacies and mutual incomparabilities in the sources made it extremely difficult to summarize clearly details of the development of production and the causes of the improved yields. Chapter 10 attempts this important task.

The information available on distribution and surplus mobilization shows that approximately 20 percent of gross paddy output was procured by the state.[5] There was little indication of appreciable cash allocations supplementary to the distribution in kind, but this was usually well over subsistence levels. Although paddy production tended to grow faster than deliveries to the state, the residual increase was often used to feed the collective pig herd rather than distributed to cooperators (see chapter 8).

The use made of the available labor in the cooperative was also impossible to quantify thoroughly. At aggregate level the only data available were those for "cooperative members of working age." Here the concepts used in the sources were as follows: the "available labor force of working age" was supplemented by "minor" labor—the young and the old. Statistically, this was expressed as the "equivalent labor force" (*lao dong da quy**). But this total figure was distinct from the labor actually mobilized ("permanent labor"), which received workpoints and could be measured in numbers, labor-days, or workpoints. Unfortunately there seemed to be a tendency to pay minor labor mobilized at harvest time via payments to their family on the "account" of more permanently mobilized workers. The situation was complicated and requires detailed interpretation, which will be postponed until chapter 7.

The sectoral allocation of labor varied greatly from cooperative to cooperative. Relatively complete data were only available for some, but the proportion of

Table 5.3

The Sample in Context:
Comparisons of Means for Selected Variables ca. 1975

	Sample	DRV Average
Population	5,200	1,050
Cultivable area (ha)	371	134[a]
Cultivable area per head of population	0.07	0.13
Rice land/ha/coop.	280[b]	65[a]
Rice land/cap.	0.06[b]	0.06[b]
Rice yield/t/ha (paddy)	6.2[c]	4.4
Production/head of population (tonnes)	0.34[c]	0.29
Collective pig herd per cooperative	660	38
per capita	0.12	0.04
Private pig herd per cooperative	1,740	340
per capita	0.30	0.34

Source: Sample calculated from various sources (see bibliography).

[a]Excludes Thang Long.

[b]Excludes Dai Thang (Quynh Luu), Thang Long, and My Tho.

[c]Excludes Dinh Cong and Giao Tan. "DRV" calculated from KTVH, tables 1, 57, and 69; also Hanoi II. The cultivated area of delta cooperatives is not directly known; it often includes "midland" (*trung du**) areas, while the data from individual cooperatives sometimes include uncultivated land. The total agricultural land use in 1974 was 2.40 m hectare (KTVH, table 1), which suggests a maximum average of 134 hectares for each of the 17,000 cooperatives (KTVH, table 68) after deducting 5 percent for the noncooperativized peasantry. No information is available on the state farm area. The rice land area (KTVH, table 67) was 1.16 m hectares (1974), giving a maximum of 65 ha/cooperative. "Rice production" is the average for 1974 and 1975.

Note: The rice yield is not the yield per sown hectare but total production per hectare of "rice land" (KTVH, table 67). While the reported average yields in the DRV were widely felt to be low by comparison with what should have been technically possible, the levels indicated were nevertheless comparatively high by traditional standards of rice cultivation in the region.

the permanent labor in the livestock branch varied between 2 and 15 percent, in the cultivation branch from 27 to 88 percent, and in the artisanal branches from 6 to 71 percent. Compared with the DRV as a whole, these cooperatives used more machinery, fertilizers, and new rice strains and had more infrastructural investment in water-works and more developed cropping patterns.

Rice production remained everywhere the most important item of staples production, and its growth was often rapid.

Differences between the cooperatives of the sample and the DRV average are hardly surprising given the presence of only one nominal cooperative in the former when probably 75 percent were of that type. Considerable evidence about nominal cooperatives comes from the histories of other cooperatives of the sample.

Limitations of the Sample and Source Materials

The primary sources upon which the sample was based have considerable short-comings. A particularly difficult point concerns the precise factors that operated to determine the balance of power within any given cooperative. Many of the main channels of influence were not recorded. For instance, it was argued above on a priori grounds that the macroeconomic environment and demographic factors would be important. There is only limited evidence, though, of the precise condition attached to priority supply of inputs to any given cooperative. And while some cooperatives in the sample came from the same districts, which may therefore be confirmed as experimental pilot districts in line for favorable treatment, there was no evidence at all to illuminate the role of the district in such matters as water access or flood control. It is known, for example, that Quynh Luu district (the birthplace of Ho Chi Minh) received considerable high-level attention. The fact that the primary materials used here do not deal with such issues tends to suggest that they were not of great importance when compared with factors specific to the internal problems of the cooperative.

Thus although the sources do throw considerable light on the socioeconomic problems of Vietnamese agriculture, their focus clearly remains that of management reform. Furthermore, the focus is on cooperatives where the local cadres made at least some attempt to implement official policy. The sources do not, unfortunately, directly examine the great majority of cooperatives where such questions were irrelevant—at least until the arrival of a reform team. This relative myopia does, however, reveal clearly the problems cadres faced, and therefore, in a highly subjective manner, an important aspect of the Agrarian Question confronting the Communist leadership. Thus the fact that the sources are addressed at North Vietnamese readers gives one direct access to real and contemporary North Vietnamese problems, which compensates for their shortcomings in coverage.

The comparisons between the sample's production conditions and those of the DRV as a whole do not show very wide differences. Some cooperatives were clearly well endowed with machines and other industrial inputs. Apart from the model cooperatives, however, these endowments were not particularly high. At the same time, the reported DRV averages were not low by comparison. Here the evidence from the poorly endowed active cooperatives is of great value. The apparent success of these cooperatives could not, generally speaking, be entirely explained by favorable outside inputs. This will be examined later in greater depth. The focus has to be on the problems encountered in mobilizing internal resources.

Finally, one crucial basic difference between all cooperatives of the sample and the DRV average remained that of size, that is, the level of amalgamation. All cooperatives of the sample were commune or supracommune level. This allowed them, therefore, to organize local interests within that old and still powerful socioadministrative unit, the commune. One of the main elements of the post–1979 policy changes was a reversal of the previous trend toward amalgamation (see chapter 12).

Because in communist-run systems the public availability of material entails a prior political decision, and because the sample used here has been determined by such availability, it is obviously not a sample in the conventional sense. Its resulting biases are, however, known with a fair degree of certainty and can be allowed for in the subsequent analysis. But in addition, the material itself has certain further defects:

1. Lack of stratification. This resulted from the relative absence of studies of nominal cooperatives and the lack of any publicly available national survey of cooperatives.[6]

2. Focus on managerial reform. There was very little information about household activity, free-market opportunities, relations with the state, and many other areas of great interest.

3. Lack of "socioanthropological" information. There was almost no information about the underlying alliances and political relations that affected dominance (see chapter 3). Accepting that they existed—especially between the different socioadministrative rural organizational levels—could not compensate for this. Perhaps the most important social phenomenon went largely unrecorded because the authors of the sources were reformist outsiders.

4. Lack of proper statistical economic data. The information in the accounts of the advanced cooperatives of the sample was not fully reproduced in the sources. It was sometimes possible to reconstruct some of these data, but this remained unsatisfactory. In some cooperatives such data were worthless.

5. Lack of full geographical information. Despite a search of available maps, it has not been possible to identify precisely the location of each cooperative (i.e., its commune as well as the district and province).

Nevertheless, such materials represent sources of great value while researchers are still faced with the impossibility of direct and unregulated observation.[7]

6

The Reform of a Nominal Cooperative—My Tho, 1974–1975

The situation in My Tho before the 1974 reform was the most extreme of all the communes covered by the available case studies.[1] Yet as a nominal cooperative it operated in ways closest to those of the great majority.[2] Its experiences during reform therefore provide an extremely valuable example of the conflicts accompanying the transition from nominal to active status.

The Pre-Reform Situation

Before amalgamation in 1973 a number of hamlet-based rice-producing brigades had acted with almost total independence. Local interests had seen to it that the commune's three cooperatives were used simply as passive intermediaries between them and the state. This situation continued for some months after amalgamation, but with the assistance of sizable financial and human resources (the latter in the form of an outside cadre team) the larger cooperative was eventually reformed. It then operated with some degree of approximation to the principles of the NMS and the Statute. When the articles describing the reform appeared in late 1974 this process had not yet been finished.[3]

Problems of Information

Before amalgamation the smaller cooperatives had exercised very little economic control over "their" resources. They were effectively dominated by their constituent brigades. As a result, the cooperatives' reports (especially those referring to production and distribution) were probably falsified.[4] This leads to considerable analytical difficulties. The non-neutrality of the sources is clear here, for the natural tendency to underreport production to reduce pressure for deliveries to the state was simply condemned out of hand. No attempt was made to examine the reasons for such behavior, partly because reformers saw the cooperatives as management units rather than tax units and therefore did not abstract from the interests involved: "The cooperatives' way of planning and accounting was usually lying and inadequate; the district never checked the figures properly" (80, p. 16).

The reported data contain some important internal inconsistencies. The commune had a population of nearly 3,000, with a cultivable area of just over 400 hectares. Reported annual paddy production during the period 1968 to 1973 was on average nearly 950 tonnes. The precise size of the working-age labor force was not given, but according to the available data[5] the productivity of land was relatively low and that of labor relatively high, with an average of 1.5 tonnes per laborer, 1.9 laborers per hectares of rice land, and a quoted "land yield" of around 3 tonnes per hectare. The commune was well endowed with land relative to its population, but the reasons for this are unknown.

On paper the cooperative had a substantial potential for increasing the existing surplus above subsistence requirements. This was because of the scope for expansion of the (minor) spring rice crop. The cooperative's records apparently maintained that the area of spring rice collectively produced in the commune (i.e., in the three smaller cooperatives before 1973 and then the larger one after amalgamation) had risen from around 5 percent of the total sown area before 1970 to nearly 48 percent in 1973. This development was in keeping with directives encouraging increases in the sown area.[6] Despite this rise in the reported sown area, however, total indicated paddy production did not rise significantly. This was explained by an apparent fall in the Xth month crop yield. But the Xth month area is not usually reduced significantly by an increase in the spring rice area (see chapter 1). The figures actually purported to show that the (minor) spring rice was yielding more than the traditional major Xth month crop. Nevertheless, "the level of output for the whole year for the years when the proportion of spring rice was higher did not surpass the total for the year 1969 when the proportion of spring rice sown was insignificant" (80, p. 13). This was obviously suspicious. The available data are given in table 6.1.

One of the most strikingly odd things about these data is that they suggest that in 1971 the Xth month crop area fell very sharply—by over one-third compared

Table 6.1

Reported Data on Output and Yields, My Tho (unit: tonnes)

	Total	Spring crop output	Xth month Output	Xth month Yield, tonnes/ha
1969	n/a	ca. 50	n/a	2.44
1970	1,072	571	501	2.52
1971	858	628	230	1.84
1972	n/a	605	n/a	1.92
1973	n/a	596	n/a	n/a

Source: Dang van Ngu, NCKT 80, passim.

with 1970. The impression given is that Xth month yields peaked in 1970 and then declined sharply once the spring rice crop had started to show substantial results. As a result, total reported output gains were made to appear surprisingly low. Despite the well over 500 tonnes gained annually from the new spring crop, reported total annual rice output in the two years 1970 and 1971 was on average not much greater than the 1968–1973 average. Thus the cooperative's apparent success in growing spring rice was not being reflected in its reported aggregate staples output.

There are a number of possible explanations for this. To deliberately take a "reformist" perspective, the most charitable explanation is that there was some form of technical problem. For instance, some input—such as fertilizer—might have been in limited supply and removed from one crop to be used on the other. Pest growth in response to the new crop could have been reducing the biological yield of the Xth month crop. But all these seem rather unlikely. The reported fall in yield for the Xth month crop between 1970 and 1971 was very great, and the total output figures suspiciously constant. The sources mentioned none of these technical difficulties.

Alternatively (and uncharitably), the figures were simply lies. Dominant interests in the collectivity did not want to raise the cooperative's liabilities to the state by reporting the increased production. Here they were in fact rather successful. The state tax had been fixed at 310 tonnes in 1970 but was only met in that year. In one year (possibly 1973) the cooperative had even managed to get 35 tonnes back out of a reduced payment of 198 tonnes. This was presumably based upon hardship arguments, which might not have worked if the cooperative had been showing a 50 percent staples output gain over the previous five years. The strong suspicion is that output was systematically underrecorded.

There are further difficulties with the reported figures for collective distribution. These suggest that a substantial proportion of collective output net of state

Table 6.2

Estimates of Collective Paddy Distribution Based upon Reported Paddy Output, My Tho

	Total pro-duction less	Estimated deliveries to the state	=	Retained output	Estimated population[a]	Estimated retained paddy output per cap. per month
1970	1,072	310		762	2,460	24 kg
1973	858	163		695	2,900	20 kg
1968–1973 avg.	950	245		705	2,735	21 kg

Source: Estimated from data in Dang van Ngu, NCKT 80.

[a]Assumed to be growing at 2.5 percent per annum.

procurement was not distributed to the cooperators in any way. It is possible to reconstruct some approximate figures to show the likely real situation. Let us assume that the highest delivery to the state was in the year of the best apparent harvest (1970), and the lowest in the worst (1973).[7] This suggests the estimates given in table 6.2.[8]

The amount of paddy distributed monthly to the cooperators was reportedly "unstable," but over the years 1968 to 1973 it was said to have averaged 16 kg with extremes of 18.6 and 14.5.[9] This is far lower than the estimates in the right-hand column of table 6.2. Even while taking the reported output figures as given, there was a large estimated gross surplus after distribution to cooperators and deliveries to the state. In addition, the reported output levels may have been gross underestimates. Without taking any account of this, the surplus amounted to just over 22 percent in 1970 and just over 27 percent in 1973. Over the six-year period 1968 to 1973 it averaged around 25 percent (see table 6.3).

This obviously does not prove that nearly 20 percent of net income was extracted and disposed of by the commune's cadres. One reason would be that they would have had to allocate around 10 percent to the funds and other areas such as feed and seed. But the output data are subject to great suspicion, and they may well have been severely underestimated. There were at the least important questions to be answered. The potential for surplus mobilization through the brigade leadership onto the free market was obvious.[10] If they had managed to hide most of the proceeds of the spring rice crop (around 500 tonnes annually) then, added to, say, half of the reported net output (around 100 tonnes), local interests would have had nearly 600 tonnes to dispose of freely each year. At 5 dong per kg this would have had a free market value of 3 m dong at a time when state officials' salaries were around 50 dong a month.

There was, unfortunately, no further information of note concerning other areas of collective economic activity before amalgamation.[11]

Table 6.3

Estimates of Gross Annual Output Distribution, 1968–1973

	Tonnes of paddy	Percent
Brigades allocated to cooperators	525	56
To state	246	26
Residual	174	18

Changes in the Management System

The focus of the sources was upon reform of the management system. They summarized the situation just before amalgamation as follows:

There was no proper and balanced system of organization and management, and no way of guaranteeing the implementation of the principle "unified economic activity, unified distribution"—on the contrary, the conditions for "work in brigades, distribution in brigades" (*an chia doi**) were created.

—The system(s) of material responsibility and economic calculation were not followed in organization, nor in the management systems, nor in the economic activity of the cooperative.

—The cadres were not directly led by their superiors to increase the level of management and techniques, and so did not know how to manage production. Competent cadres did not develop their good points, incompetents were not shown up and judged in good time. Because of this the cooperators lost ideology,[12] had no peace of mind, no enthusiasm for work, mistrusted and feared cadres, and did not dare to assert their rights to mastery.

—None of the organizations of the state had yet realized their responsibilities, with regard to both protecting the cooperative and seriously carrying out the economic contracts and the systems of policies concerned.

—Because of this various technical advances, things that have become normal in other places, still have not occurred in My Tho. (80, pp. 15–16)

The "proper and balanced" system of organization and management should have resulted in a cooperative plan. In addition, the Management Committee should have added to this a proper knowledge of developments inside the brigades. Not only had this not happened, but in practice quite different methods had been used to regulate the operation of the collective structures that higher levels required the collectivity to maintain. This was quite characteristic of nominalization:

The job of fixing the plan of productivity and production rested upon fixing the distribution of the agricultural tax (among the brigades) and then increasing it by a proportion so as to arrive at the level of production to be put in the so-called plan.[13] Because of this the cooperative did not agree with [go along with] the system of three contracts for brigades. The contracts given out were pieces of blank paper and had nothing to do with any system of norms. This meant that the three contracts became the basis for working [according to the system] "work in brigades, distribution in brigades" where the brigades operated by themselves according to their own ideas, and lived according to their own results, after subtracting something to contribute to paying off the activity of the cooperative. (80, p. 15)

The brigades were therefore using the cooperative as a conduit for delivering taxes to higher levels. The ephemeral plan had simply become part of this arrangement. The brigades were the focus of local authority:

The management did nothing but act as representatives of the brigades in dealing with the state in the work of buying raw materials, borrowing money, "appealing," mobilizing [e.g., troops, social labor[14]] and *supervising vaguely the encouragement of brigades* to raise the proportion of new seeds, transplant at the correct time and so on. . . . Individualism and localism are very strong among the cadres of the cooperative and of the brigades.

The plan was constructed to obtain the approval of superior levels and to [provide a basis to] ask for supplies of raw materials and loans. (80, pp. 15–16; emphasis added)

The Management Committee naturally did not know the most basic things about its "own" resources: "The cooperative did not comprehend its concrete position with regard to [labor, land, and capital]" (80, p. 15). Which said, it is not surprising that official data on the earlier period were so unreliable and inconsistent.

These quotes show clearly the situation confronting the reformers in 1974. Working in a highly nominalized cooperative, they had to establish the new Management Committee's effective control over the factors of production available to the new amalgamated cooperative. They did not inherit this control from the earlier, smaller cooperatives. The formal property rights of the new cooperative were profoundly "immanent" (see chapter 3), and doing something about it meant tackling entrenched interests. This was not easy.

The Brigades During Managerial Reform

The brigades' de facto property rights were well established. Land was scattered, with each brigade having "near land, far land; poor land, rich land" (*co gan, co* *xa, co tot co xau**) (80, p. 16). In addition labor was, as expected, fixed in brigades corresponding to hamlets. This had a strong effect upon the brigades' landholdings: "The amounts of land belonging to each brigade are extremely varied, because they result from the [division by] hamlets [in the commune]." The brigades also controlled other important means of production: "The brigades freely buy means of production, from small spare parts up to water buffalo. They also have their own funds. Only in the case of materials supplied by the state are the brigades forced to go through the Management Committee."

Control over output distribution was a natural consequence of this independence. The brigade leaders possessed considerable power: "The proportion of paddy allocated to households for drying is too high, but the difference between households is also too great—there are even some who are allocated none at all *because they have angered a cadre*—which leads to a seriously incorrect distribution" (emphasis added).

But although there was no planning and no proper system of norms, the brigades do not seem to have abandoned collective production. There was no record of subcontracting to individual family units: "The organization of labor is according to convenience, with no proper division into small groups or contract work; work which nobody wants to do leads to daily pay or increased workpoints." The evidence therefore suggests that the brigade leaders were in a strong position. They were in the habit of simply "signing the production papers, and not actually checking the amounts."[15]

The Process of Reform

In this nominalized cooperative understanding the reform process is of key relevance to the argument of this book. The sources reveal much about the way in which this happened. Here an outside cadre team was the main stimulus for changes in the cooperative's management system.

Agency of Reform

The stimulus for reform came from the district rather than the party committee in the commune. The relationship between party and cooperative was said to be "vague":[16] "The party was usually absorbed in replacing the cooperative in organizing economic activity. When a job was done wrong but in accordance with local interests and using a resolution of the collective, it was very hard for the superior level to maintain control" (80, p. 16). Nevertheless, this control was much needed: "Up until recently many groups of villagers had informed the Management Committee . . . about frequent incidents of embezzlement by cadres, but these were not quickly dealt with" (83, p. 28).

Sources in Hanoi, it will be recalled, asserted that the local party hierarchy would typically act to cope with such problems (see chapter 2). The source used here, however, asserts that such an intervention by the district was a new phenomenon:

> In the case of My Tho, a cooperative where an experiment in management reform is occurring, the permanent committee of the district is both capable of, and indeed carried out, a direct intervention to resolve the various problems discussed above. These relate to the relations between the party in the commune and the cooperative, and between the various organs of the state and the cooperative, with regard to both the execution of the new management system of the Management Committee and its chairman, and the control of the cooperators and their right to collective mastery. (83, p. 28)

This suggests that leading party members in the commune were entrenched in their "fiefs"—the brigades—and so the district had to bypass the local party network in order to push through the reforms. It was not made clear whether the Party Committee or some state organ at the level of the district was responsible. The wider political implications of this are obvious, for the source is implicitly criticizing the North Vietnamese system's ability to regulate grass-roots party organs.

The sources analyzed the process of reform by considering three different areas in turn. These were:

1. The reorganization of production. In practice, this meant choosing the cooperative's output pattern, and the corresponding system of specialization by both brigade and branch.

2. Reorganization of the management structure of the cooperative and establishing basic management principles.

3. The construction of a new management system. By this the sources primarily referred to such management functions as planning, control of labor and product, distribution, and so on.

This way of approaching the task of reforming the nominalized cooperative was quite consistent with the reformers' understandable attitude that managerial methods were the major issue. From the point of view of those about to be reformed, however, things were slightly different.

Conflict over Resources

The main problem was in fact the difficulty encountered in establishing the Management Committee's rights to control economic resources. This was reported as resistance to the reform process coupled with persistent localism and other "irrationalities."

When the commune-level cooperative was set up, the resources of the smaller cooperatives had to be transferred formally to it: land and labor remained for the moment controlled by the brigades, so conflict centered instead upon other areas—rice-stores, money, and so on:

The property, above all the financial resources and the seeds for the May 1974 [planting], was sold by the three cooperatives to their successor. This only happened after a long period of violent struggle against feelings of partiality [localism], individualism, active scattering of property, and successful attempts to take advantage of the moment when the organization was not yet secure to get hold of the collective's property. (One cooperative took 9 tonnes of paddy allocated for livestock-rearing that should have been sold to My Tho, "selling" all the paddy in the public store house and putting it into the hands of a number of principal cadres and families.) (80, pp. 14–15)[17]

Thus when they arrived in 1974 the reformers had to deal with a conflict-ridden and recently amalgamated cooperative. Eventually, however, the new Management Committee did manage to bring the brigades under control and drastically curb their powers. Once this had happened, "The responsibilities of the brigades [now] are to create products and the value of products; this does not include bearing responsibility for and having the right to supply means of production, to distribute [the product], consume products, and realize their value" (81, p. 28).

At the time the sources were written, the reform process was still unfinished. There had originally been thirteen rice-producing brigades whose landholdings varied between fourteen and fifteen hectares. But although there was every intention to reorganize these so as to have fifty to seventy workers and twenty-five to thirty hectares in each brigade, this did not seem yet to have happened. The basis for these numbers was meant to be the capacity of the brigade leadership, identified as the brigade head and his deputy. The new Management Committee had, however, taken certain large means of production (such as a tractor) away from them "until they had a high enough technical level." In keeping with the provisions of the Statute (see chapter 2) they had also put one of their number, a deputy manager and member of the Management Committee, in overall charge of the rice-producing brigades: "These [rice-producing] brigades set up, under direct management, . . . the work for each day according to the plan of the deputy manager [responsible for] cultivation" (81, p. 30).

At the time the articles were written, land had not yet been reallocated among the brigades. This suggests that the reform was still far from complete. Despite this, however, the Management Committee's position was now far stronger, especially vis à vis the brigades.

Management Committee Dominance

Under the new system power had become far more centralized, and members of the Management Committee were able to exert considerable authority. A key element of this strategy was the use of the cooperative's production plan as a means of controlling the brigades. This was part of a wide battery of measures, however. For instance, cooperators now had to have the signature of a deputy manager before they could open the cooperative's stores. Such detailed controls required a corresponding growth in the cooperative's bureaucratic apparatus. To cope with such work the accounting office was expanded until it contained twelve people—four responsible for the aggregate accounts, including the state accountant, and eight "keeping an eye upon the brigades" (p. 29).

In the face of such intense Management Committee pressure the brigade leadership's attitude improved rapidly. It was reported that the "most rapid advances" in the ideology of commune cadres occurred among the brigade leaders of the cultivation branch (82, p. 30). It is of interest that during this shift in the local power balance the Management Committee seems to have shown some willingness to compromise. The committee perhaps acknowledged the brigades' crucial functions in the control of production and output mobilization. Possibly in recognition of this, a brigade leader who had underfulfilled his part of the cooperative's plan on a large area was not penalized according to the newly established system of penalties and bonuses (83, p. 24). More concrete measures helped encourage obedience, and the cooperative now paid all cadres enough for them not to have to do manual work. This helped to separate their interests from those of the cooperators in their brigade.

Yet the most striking example of persistent localism, in this case at village level, was on the part of a deputy manager:[18] "[There are still many problems] above all in contractual relations, where the heavy weight of localism, individualism, and 'convenience' persists; for instance the deputy manager responsible for cultivation did not carry out properly the order of the manager to allocate a buffalo to a brigade in another village from the one the deputy manager lived in" (82, p. 30).

The overall level of brigade specialization remained very low. This was justified by the absence of modern tools: "specialization of brigades should arise when (a) a high level of technology required direct supervision, or (b) direct supervision is needed, for any other reason, or (c) there were technological reasons—tools, for instance" (81, p. 28). A new water brigade ran the pumps and a technical brigade was responsible for sprouting seeds, developing techniques generally, spraying against pests and diseases, and acting as a reservoir of mobilizable labor. There was no collective pig herd; funds were "not sufficient" to allow pigs to be bought.[19] There were, however, brigades for rearing fish, raising ducks, making bricks, carpentry, and mortar, and also for basic and semispecialized irrigation. The Management Committee used the latter for absorbing the rice brigades' "surplus" labor during slack periods. The others were still very much in embryo. A transport brigade for portering used bullock carts, while a number of work groups "fulfilled functions normally carried out by the offices of the Management Committee." For instance, one group was responsible for the storehouses, and another for buying and selling (81, p. 31).

The level of specialization was in fact rather low, especially compared with model cooperatives. Some of the cooperative's specialized subunits only existed on paper. As an indication of this, it was noted that certain brigades did not yet have leaders. But the brigades now enforced a far higher level of labor discipline, and ensured the Management Committee a better and more effective supply of labor. Downward pressure on rates of remuneration was confirmed by a report showing that the brigades could now guarantee work quality without paying out extra workpoints:

> The leader of brigade B obliged the cooperators to pull up the rice seedlings twice and replant it again in order to get it done at the correct density, but with no extra allocation of workpoints. The leader of brigade C penalized cooperators who deliberately did not carry out their work properly: they had brought out seedlings that should have been sown on 8 sao, but in reality only sowed 4 sao. (83, p. 25)

Production and Distribution

Important changes were made to the system of distribution. The new cooperative had apparently faced a chronic shortage of working capital. Since the three smaller cooperatives had apparently done all they could to obtain loans, the new My Tho inherited substantial debts.[20] The existence of this financial obligation seems to have helped reformers to justify an increased level of deliveries of real economic resources to the state. In practice, the reform team arranged for mobilization of an increased rice surplus sometime in 1974. This took the form of a sale to the state at an "incentive price" whose proceeds were used to pay off the cooperative's debts.[21] Unfortunately, the available data are again far from straightforward to analyze.

During the reform there is some evidence that the cooperative received a massive injection of funds. Over an (unstated) period of less than twelve months the reformist Management Committee of the new cooperative was allowed to borrow over 0.8 m dong. Such a sum was approximately equal to the yearly gross accounting output of a relatively advanced cooperative and at the then state duty-price of 0.32 dong/kg equivalent to over 2,500 tonnes of paddy. It is hard to discover the real counterpart of this financial flow, although there is evidence of an allocation of chemical fertilizer.[22] This may help explain part of the sharp jump in reported annual production compared with the previous half-dozen years. But it is not likely to explain all of the increase. If the suspicions about output diversion by the prereform cadres are correct,[23] the reported output jump indicates a very high level of earlier underreporting, for table 6.4 shows that the increase was over 50 percent.

The Management Committee sold part of the reported output increase to the state at a price estimated at 1.4 dong/kg, which was favorable to the cooperative

Table 6.4

Increases in Reported Production During Managerial Reform, My Tho

	Production	Area	Implied yield
1968–1973	950 tonnes	315 ha	3 tonnes/ha
Harvest year[a]			
1973–74	1,450 tonnes	363 ha	4 tonnes/ha
Spring crop	440 tonnes	146 ha	3 tonnes/ha
Xth month crop	1,010 tonnes	217 ha	5 tonnes/ha

Source: Dang van Ngu, NCKT.

[a]The source simply refers to "1973–74" without explanation. Note that the spring rice crop is harvested around June while the major Xth month crop comes in around December. Thus it is not clear precisely which crops are referred to here. It seems most likely that the reform team went in very early in 1974 and therefore quickly got control of the just harvested Xth month crop. In the report, this was then included in "1973–74." This would explain the sharp jump in reported Xth month output. The spring crop referred to could be either that of mid-1973 (preform) or that of mid-1974. Note that there was no dispute over the size of the spring rice crop.

Table 6.5

Estimates of Physical Allocation of 1973–74 Rice Harvests (tonnes of paddy)

Total production	1,450
To funds	65
Sold to state at high price	250
Distribution according to labor	750
Ordinary deliveries to state	310
Residual[a]	75
Estimated share of total production delivered to state	39%

[a]Note that at 100 kg per month the residual 75 tonnes if distributed to cadres would have been sufficient for sixty-two. Twelve cadres have already been identified in the Accounting Department, while there were at least twenty brigade heads and deputies. No mention was made of a priority distribution to needy families.

because it was well above the 0.90 dong/kg that was the normal incentive price.[24] This sale of 250 tonnes paid off 180,000 dong of the outstanding debts. There is, however, no complete breakdown of the uses to which the 1973–74 harvest (see note to table 6.4) was put. Funds and paddy were allocated to some of the cooperative's funds (64 tonnes and 21,000 dong[25]), while the Management Committee paid increased salaries to cadres. Such personnel had previously not earned enough to enable them to devote all their time to organizational work, but in 1974 they received the following: manager—665 marks, equivalent to 55 dong plus 110 kg per month; deputy manager—595 marks, i.e., 50 dong plus 100 kg; brigade heads and deputies—400–560 marks. Other cooperative cadres were paid from 350 to 525 marks. The average cooperator earned about 350–390 marks, or about half the amount paid to the manager.[26]

The Management Committee raised sharply cooperators' average monthly paddy distribution for work amounted to 21 kg. This implied a total for the entire cooperative of around 750 tonnes, so a possible reconstruction of the rice harvest allocation for the 1973–74 harvest year would be as shown in table 6.5.

This percentage share was, of course, well above DRV averages (see chapter 1). The sources gave two justifications for this sharp increase in the mobilized surplus. One was the need to pay back money borrowed from the state, and the authorities in fact encouraged this by paying a substantial price premium. The other came from the powerful ideological notion that there was a normative upper limit to the level of rice distributed to cooperators.[27]

> If all the . . . funds have been dealt with—(1) fixed reserves for paying taxes, making duty sales, providing seeds etc. and (2) reserves for livestock rearing— but the level of staples available to the population is on average more than 17 kg per head per month, then more must be allocated for sales at a high price to the state. In present conditions if My Tho did not sell at a high price then it would not be able to meet its needs for capital, in other words simple reproduction would not be possible,[28] not to mention expanded reproduction. (83, p. 22)

In many ways the sources remain frustratingly opaque, yet the most believable picture that emerges from all this is of conflict between the old and new management teams over the proceeds of the new spring rice harvest. Previously this had gone unrecorded, and most of it was probably sold onto the free market. After reform the new Management Committee delivered a large proportion of it to the state in the form of a fulfillment of the tax obligation of 310 tonnes plus the "incentive price" sale that helped repay the cooperative's debts.

The Plan, Managerial Control, and Distribution

The authorities intended that a cooperative should base its plan upon a system of norms that made explicit the relationships between inputs and outputs. This "scientific" method purported to give a determinate relationship between inputs and outputs that could actually be realized as the plan was implemented. The basic problem with this way of thinking remains the instability and unpredictability of the arithmetical relationships involved, especially in tropical agriculture. Furthermore, this system could never be all-embracing because exchange relations persisted in certain areas, most importantly the household sector. Nevertheless, the system of norms could be a fundamental part of the socioeconomic control mechanism. Properly used, it allowed the Management Committee to monitor the brigades' production; it was the basic planning tool for cooperative-wide resource allocation, and it effectively facilitated centralized distribution by providing a means of ensuring that the value of the workpoint was the same throughout the cooperative. Although this had to mean a relative increase in the value of the workpoint in the poorer brigades, which might therefore support the NMS, this was never mentioned. Nor was the possibility that overall output gains and corresponding increases in collective incomes would compensate the richer brigades. These significant absences from the sources suggest that other factors such as the overall shift in the local power balance were more important. For instance, better labor control would probably make it easier to enforce the restrictions on "own-account" activity that usually accompanied the NMS.

The content of the reforms in My Tho was classified above in terms of the reorganization of production, production units, and administration. The underlying sense of a drive for control and authority is suggested by the sources' comment that the reforms meant, in essence, answering the questions: "What work is to be done? By whom? How—with respect to technique, economic and organization? And if it is done well, what happens?" (81, p. 25).

A "correct" planning procedure resulted in the General Assembly's issuing an "Instruction" to the Management Committee. In the case of My Tho, this contained:

> The outputs of all branches: the value of total production calculated according to consumption [market?] prices. The value of consumption goods calculated at duty-prices. The value of consumer goods calculated at supra-duty prices: production costs and consumption costs of each product; the total number of work points for the administration; the value of the work-day—including all payments in kind—and the average level of income; the value of all basic construction at cost-prices; profits and distribution of finances. (81, p. 31)

The manager then handed this on to the brigade heads.[29] The brigades, as holders of economic resources, had their own production requirements from which the value of production could be calculated using the brigade's own cost prices. Full details of this procedure were not available, but preliminary estimates of labor incomes could be made, such as "the value of the average workday of the brigade members" (82, p. 31).[30]

Under the reformed system the brigade leadership and the Management Committee reportedly constructed the plan together before submitting it for the Assembly's approval. The brigade head was reponsible for computing balances[31] (of which there were three or four[32]) and tables covering seeds, land area, labor, fertilizer, green foodstuffs, livestock, the need for maintenance (threshers, repairs to tools), raw materials, and so forth. These were probably simple material balances. But he did not arrange contracts with other brigades. The Management Committee thereby retained close control over horizontal relations between the cooperative's subunits, for instance those between irrigation and cultivation brigades. The Management Committee only issued the contracts governing such relations after the manager had personally checked them: "The brigade head must himself set up the balance tables for materials and labor *in accordance with the amount of work allocated* [to his brigade]. It is only with regard to materials demanded of the cooperative that he must get the manager's approval" (81, p. 32; emphasis added).

In My Tho planning was said to have gone beyond the III-point contract system:

> This system of planning has replaced the III-point contract system and resolved the contractual relationships between each organization by means of the specific duties given out to them; planning guarantees [the successful results of] instructions from the Management Committee with regard to all economic activity from production to finance, from supply to end-use. (81, p. 34)

Apparently this process only applied to the cultivation brigades. As has been seen, most of the others were either in embryo or directly managed by the Management Committee.

The value of the mark[33] had been stabilized at 2 kg "based upon the plan"; they were attempting to keep monthly payments stable, which implied that there was no post harvest distribution of the crop to cooperators (a phenomenon observable elsewhere), although there may have been one to brigades. There was "unified distribution" within the cooperative and the Management Committee's control had greatly increased, but the workpoint system was not universal. Supplies of fertilizer from the privately owned pigs were paid for directly in paddy. These sales were obligatory, but "carrying out these duties well" [*sic*] would see the supply of staples to a household rise by one-twelfth (83, p. 23).

The new management team allowed the cooperators greater participation in decision making:

> [Since the cooperative was set up, they had had three meetings] in order to approve the plan, the cooperative's own Statute, the system of management and the technico-economic norms; the cooperators' representatives had taken a first step in discussing and resolving problems properly, and not just agreeing to everything as previously. For instance the Assembly determined the level of output, the value of the mark, the level of foodstuffs, the cadres' incomes, the structure of the cooperative's management system, but did not approve the level of manure contributable that was suggested by the Management Committee. (82, p. 30)

There was, unfortunately, no further information on the role of the ordinary cooperators in the political conflicts that accompanied managerial reform.

Conclusions

In 1974 the cooperative was still in the process of reform, and much of the discussion in the articles dealt with intentions rather than fully implemented changes. The basic pattern was clear, however; in addition, some distinctive features of My Tho were apparent.

The main thrust of the reform process was the Management Committee's progressive domination of the brigades. Initially, this took the form of a struggle to implement the new cooperative's property and control rights. This extended to cover collective product, means of production, land, and labor. The reformers had to ensure that resources were properly transferred to the new cooperative and the position then secured by setting up an enlarged bureaucratic apparatus. This centered on the accounting office and was backed up by administrative measures such as countersigning. Although the new Management Committee had not yet reallocated land, its use of certain specialist brigades to utilize "surplus" labor was the beginning of a cooperative-wide system of labor allocation. The jump in reported production strongly suggests that surplus output previously retained improperly within the cooperative (most probably by the brigade leaders) was now effectively controlled by the cooperative's reformed Management Committee.[34] The brigade leadership became subject to the planning process. It was given an output target and allowed access to certain means of production with which to attain it. As part of the control mechanism, the output target contained certain defined tasks that it was the brigade's responsibility to carry out. As this was done, records were kept in order to justify labor expenditure and so allow the brigade's members to claim remuneration from the cooperative for the work-points earned. Since the volume of paddy threshed, reaped, and carried was therefore clearly documented, it became almost impossible for the brigades to retain their previous control over output. This allowed the Management Committee to reallocate resources both within and beyond the boundaries of the cooperative.

The apparent absence of the III-point contract system was a distinctive feature of the reformed My Tho. The new Management Committee replaced it with close supervision of brigade activity by individual committee members. Precisely why this was so was unclear, but it may simply have reflected the relative weakness of the reform team in the face of local cadres, given the latter's contacts (if not collusion) with elements at the district level. Regulatory agencies had not been carrying out their responsibilities properly ("the district never checked the accounts properly"), and the reform team's position outside the main channels of regulation, both of party and of state, was apparent.

My Tho has great value as an illustrative example. This is primarily because it confirmed the existence of behavior that a priori considerations had suggested would be present during the transition from nominal to advanced status. It supported a number of basic theses and above all made it quite clear that North Vietnamese communes could be areas of sharp and often overt conflict. The reform process in My Tho showed the fundamental importance of the collectivity's tripartite division into brigades, Management Committee, and households. Conflicts between the first two came to the fore as the cooperative was "made to work"; this meant that there was now the prescribed "unified economic activity and unified distribution" at the level of the cooperative. And here, in particular, could be seen how the NMS was adapted, with brigade independence further reduced by the abolition of the III-point contract system in favor of something closer to direct management.

It is argued here that while an effective and powerful Management Committee was a major requisite for development of the collective economy in these cooperatives, full-scale implementation of the NMS itself was not. The NMS may have helped a Management Committee to hold its position, but where this was particularly weak (for other reasons) the NMS was often adapted in order to strengthen further the committee's position. This happened in My Tho. Conversely, later chapters will show how other management committees "liberalized" the NMS when they were clearly strong enough to use it in its pure form if they so wished. The best example of this was Van Tien cooperative.

A basic question concerned the relationship between the development of production and the production organization. In My Tho the reformed Management Committee's control of brigade cadres, and indirectly therefore of ordinary cooperators, was based predominantly upon the authority of the incoming cadre team and the use of administrative measures to control product. Because no real process of accumulation and production development yet existed, neither the use of new technology nor the fine penetration of the production process accompanying diversification and specialization was important. Despite the sharp increases in reported output, reform in My Tho essentially dealt with a relatively static economic system. Examining in practice the factors influencing the transition from nominal to advanced status listed in chapter 3, the most important was probably the presence (in a surplus-producing commune) of a team of outside cadres. Other factors—especially the availability of nonfinancial resources from the state and incentives to cooperators—were not apparently important.[35] But these materials only referred to the first stage of reform. Later sources indicated a substantial real development of production which may have helped sustain the changes brought in by the reformers.[36]

7

The Cooperative and Its Constituent Households— The Labor Mobilization Problem

The next four chapters discuss examples of economic behavior drawn from the experiences of the sample. This chapter concentrates upon issues of labor mobilization and associated conflicts in cooperative-household relations. The next two chapters consider further examples of such conflicts in the pig-rearing and subsidiary branches. Finally, chapter 10 discusses factors influencing technical innovation and other changes accompanying improved rice yields.

This study argues that social authority played an important role in determining access to economic resources, and therefore in the conditions governing the introduction of new methods of production and the possibility of increased output and surplus mobilization. Such economic factors as control over production and output distribution were not the only sources of social authority in a collectivized agriculture such as that of North Vietnam, but they were probably very important.

Economic sources of social authority (see chapters 2 and 3) differed at Management Committee, brigade, and cooperator levels. These differences provided a basis for conflicts over access to means of production, control over labor and the production process, and output distribution. The My Tho case study provided strong empirical support for the postulated existence of so-called nominal cooperatives where the statutory rights of the cooperative's Management Committee were not generally respected. The transition to a status close to that prescribed by the NMS appeared to require the successful domination of the brigades by the central managerial organ of the cooperative—the Management Committee.

Hypotheses

Attempts at managerial reform in North Vietnamese cooperatives usually had to overcome certain constraints. With regard to the cooperative's environment, these were primarily the poor incentives to participation in the collective sector that resulted from the high free-market price levels and the low levels of supplies of means of production from the state. Inside the collectivity, the potential alliance between cooperators and brigade leaders resulted from their shared interest in colluding to reduce procurement levels and labor mobilization, defend privatization of cooperative assets, and so forth. But reformers benefited from the substantial possibilities for improved resource mobilization. Even nominal cooperatives needed to maintain some level of collective production, and this was often very inefficient. Reform could therefore raise the economic efficiency of the collectivized part of the collectivity solely through improved utilization of existing resources within the collective sector. But a cooperative's higher output could also result from increased mobilization of resources obtained from the private sector. Because of this, large yield increases on collectively worked land were frequently possible in the short term without substantially increased inputs from outside the cooperative.

A Management Committee did not have to use administrative measures to obtain resources from the private sector. If average output within the collectivity rose as a result of higher collective output, and if output in the private sector did not fall, and if the state did not raise procurement, then local interests might have accepted such a strategy. If so, then the Management Committee would be able to use mutually beneficial exchange as a basis for its relations with the private sector. Put simply, the committee would pay (perhaps with a bundle of commodities and services) for the things the cooperators supplied to it, rather than relying upon the imposition of duties.

This suggests that active cooperatives with management teams whose social authority was strong could often appear to tolerate a far greater degree of brigade independence and decentralization than those who needed to bolster their weak position by imposing the entire NMS apparatus. Van Tien cooperative was an outstanding example of the former. Dai Thang (Xuan Thuy) provided an example of a cooperative with a weak Management Committee, while Hai Nam and Giao Tan cooperatives showed some remnants of nominal structures.

The important role played by control over labor in conferring social authority, coupled with the ideological position, suggests that the sources would often

Table 7.1

Labor Mobilization Problems in Cooperative-Household Relations: Summary of Main Findings

Information issue	Time-series	Cross-section
Labor utilization	Apparent large changes (declines) in some measures of labor-use in some cooperatives.	Increase in labor mobilization as management strengthens. "Outsiders' " opinions that labor mobilization problems rise as "own-account" activity increases.
Distribution of collective output	—	"Brigade-level" system in weakly managed cooperatives; "cooperative-level" system in strongly managed cooperatives. Mean level of distribution of staples from collective exceeded subsistence. Severity of penalty-bonus systems related to strength of management.
Labor norms in rice production	—	Inadequate data to explain differences. According to most cooperatives' norms, total labor inputs needed to produce staples required to reproduce labor force usually quite low.
Deliveries of agricultural produce to state	Rapid increases in some cooperatives.	Generally higher proportion of gross output in well-developed and strongly managed cooperatives.

discount the importance of the household sector to the cooperators. The sources were unlikely to present a balanced view of the relative returns to cooperators of work in the two sectors, and they would probably undervalue the economic results of "own-account" activity. But the difficulties involved in mobilizing labor into the collective sector largely resulted from the cooperators' comparison of the magnitude and stability of returns from the collective and private sectors. It follows that suprasubsistence levels of collective distribution should not necessarily result in fewer problems in mobilizing collective labor. Collective distribution was well above subsistence in many of the cooperatives of the sample.

The basic differences between the nominal and other types of cooperative corresponded to two distinct types of distribution system. This was mentioned in chapter 3 as a comparatively direct reflection of the balance of social authority between brigades and Management Committee, at least insofar as this had an economic basis. Cooperatives with strong central managements were expected to have established their control over distribution and abolished brigade-level distribution.

The following discussion uses abbreviations to refer to the cooperative type represented by the particular example. These are:

M   — model (Vu Thang and Dinh Cong)
N   — nominal (pre-reform My Tho)
AWM— active, weakly managed (Hai Nam, Giao Tan, Dai Thang [Xuan Thuy])
ASM — active, strongly managed (Dai Thang [Quynh Luu], Van Tien,
         Thang Long)
A   — active (post-reform My Tho).

Table 7.1 summarizes the main findings.

Basic Issues of Labor Use

A cooperative's labor supply measured in labor-days was usually well in excess of the specified norms of rice production. Even if some allowance is made for risk, and especially the periodic need for much work to control water (primarily because of flooding), for parts of the year collective rice production would probably not have suffered from the absence of much of the labor force. Table 7.2 presents data on labor availability and use for the sample. Because of household sector employment, however, total production in the collectivity was not nearly so insensitive to labor withdrawal. The sources continually focus upon the problem of "surplus" labor yet ignore this issue. Attempts by cadres to absorb this "superfluous labor" (*lao dong roi rao**) encountered opposition precisely because cooperators were busy on their own account.

These issues became more important if the management team was divided by conflict between the brigade leadership and the Management Committee. Conversely, in the better-managed "advanced" cooperatives the phenomenon of labor "running away to the market" (*di chay hang xen**) was less of an issue. Exacerbating such problems was any lack of control of the "5 percent" land. Cooperators not only legally supplemented these with the private "gardens and plots" (see chapter 2), but also frequently managed to extend them until they covered well in excess of 5 percent of the cooperative's land:

> An objective factor makes it hard for the cooperative to organize labor: the cooperative is near a market, and the phenomenon of "run to the market" is very widespread. The 5 percent land is still too extensive in a number of brigades. The cooperative needs to find some way of refixing the level of 5 percent land of the family economy to ensure that the distribution system is fair, and to *reinforce the relationship between labor and collective production*. This is really why the custom of production "at convenience" and without planning—a remnant of small-scale production—is the basic impediment to the planned scientific man-

Table 7.2

Labor Mobilization in Some Cooperatives of the Sample

	Availability[a]	Use[b]
Dinh Cong (M)		
Total		
Numbers	1,278	796 (62%)
Labor days[c]	332,000	n/a
Rice		
Numbers	n/a	n/a
Labor days	n/a	ca. 300,000
Dai Thang (Quynh Luu) (ASM)		
Total		
Numbers	1,454	900 (62%)
Labor days[d]	410,000	n/a
Rice		
Numbers	n/a	below 94
Labor days	n/a	311,000
Van Tien (ASM)		
Total		
Numbers	2,348	2,158 (92%)
Labor days[e]	610,000	379,972 (62%)
Rice		
Numbers	n/a	below 583
Labor days	n/a	ca. 200,000
Dai Thang (Xuan Thuy) (AWM)		
Total		
Numbers	3,160	2,125 (67%)
Labor days[f]	758,000	346,000 (46%)
Rice		
Numbers	n/a	n/a
Labor days	n/a	n/a
Hai Nam (AWM)		
Total		
Numbers	2,343	n/a
Labor days[c]	610,000	n/a
Rice[g]		
Numbers	n/a	(755 in cultivation)
Labor days	n/a	180,000

Sources: Derived from a variety of the sources used to construct the sample, with many of the figures estimated from indirect references. See bibliography, table B1.

[a]Understood to be those liable to labor duties.
[b]Labor identified as working for the cooperative.
[c]Labor duty estimated at 260 days annually.
[d]Labor duty approximately 280 days annually.
[e]Labor duty was 260 days annually.
[f]Labor duty was 240 days annually.
[g]It is probable that Hai Nam's high level of labor productivity in rice production at least partly reflects poor accounting. The cooperative was not an efficient producer.

agement of labor. The work of reforming small peasants, reforming completely their spirit and customs, is a task that will take many generations.

"With regard to the small-scale peasantry—only with a material-technical basis of tractors . . . and electrification on a large scale will it be possible to solve that problem" (Lenin) . . . [so] . . . if it wants to have a large output [or: scale of production] the cooperative must have the capital and a certain level of activity; above all, it must reorganize production along the lines of mechanization. If it wants to have the capital to reform production it must continuously mobilize and encourage all sources of productive labor, and manage and organize this source of labor rationally in order to make labor productivity rise.[1]

But in Van Tien (ASM) where the Management Committee's authority was apparently strong, "excess" 5 percent land was tolerated. Cooperators had simply to pay rents for it. This cooperative had mobilized a relatively large proportion of the labor force by operating with a high degree of brigade independence: "If any household has excess 5 percent land then for each sao it has to sell 120 kg of paddy a year, and for each sao of pond 20 kg."[2]

This suggested that if a Management Committee was, for whatever reason, in a position to ensure that its directives were carried out, it did not have to follow the prescriptions of the NMS regarding a complete and centralized control of economic activity. If so, it could still be comparatively successful in terms of production development and surplus mobilization. Under such conditions, privatization of collective assets did not necessarily imply weak management.

The drive to mobilize collective labor could confront a Management Committee with a dilemma. If, on the one hand, there were always productive uses to which such labor could be put, then, on the other, excess numbers appeared to make it hard to realize adequate control over production and product quality. The basic monocultural brigades with static production methods typical of nominal cooperatives were well-suited to labor sharing. Cooperators' apparent desire for labor sharing appeared to conflict with efficiency demands in collective labor:

> In a situation where the cooperative has not yet completely resolved the problem of stabilizing conditions for the workers, it is as a result necessary to share out the available work (no matter how small) in order to guarantee an income for all . . . [but] . . . this results in the cooperative's being unable either to mobilize the hidden potential of the workers or to encourage the mobilization of labor, all of which has an influence upon the productivity of labor in the cooperative [Dai Thang, Xuan Thuy, AWM].[3]

The basic tool for mobilizing labor was the posting of labor-duties. Management Committees usually stipulated the number of days cooperators had to work annually for the collective. This varied from 288 days (equivalent to between 316 and 362 labor-days) in Dai Thang (Quynh Luu, ASM) to 240 days ("neither

Table 7.3

Gross Deliveries to the State and Collective Distribution to Cooperators (paddy tonnes)

Cooperative	Date	Distribution (kg per cap per month)	Total	Gross deliveries to state (%)	Residual output	Total
Vu Thang (M)	1974	28	1,275	535 (24)	440[a]	2,250
	1975	16	730	450 (28)	400[a]	1,580
Dai Thang (Quynh Luu) (ASM)	1976	22.5	780	>420[b] (>21)	<785	1,985
Van Tien (ASM)	1974	17	1,060	160 (12)	150	1,370
Thang Long (ASM)	1977[c]	23	1,275	425 (25)	nil	1,700
Giao Tan AWM)	1974	22	1,320	120[d] (8)	60	1,500
Dai Thang (Xuan Thuy) AWM)	1974–75	17[e]	<1,700	n/a	n/a	2,100
My Tho (A)	1974	21	750	560 (39)	140	1,450

[a]There was a very large distribution of paddy to the pigs (see chapter 4).
[b]The source maintained that the "duty level of 420 tonnes was always surpassed."
[c]The data here are "staples."
[d]Here the data are the value of deliveries in dong expressed in paddy by dividing by the duty-price.
[e]The level of distribution was "less than the subsistence level."

carried out strictly nor with discipline") in Dai Thang (Xuan Thuy).

Data on labor duties were an extremely misleading guide to labor use in a cooperative because such duties were almost never fulfilled. There were also large differences between a cooperative's labor supply, measured in terms of its membership, and its "employed" or active labor force, generally known as its "permanent" labor force. This was partly because of people who belonged to families registered with the cooperative and did not submit to its management team. In addition, cooperatives often supplemented "permanent" labor by other workers at periods of peak labor demand. Such additional labor seems to have been rewarded by payments to a family member active in the "permanent" labor force.

Incomes and Distribution Systems

Absolute Income Levels and the Distribution of Collective Output

Table 7.3 summarizes the available information on collective distribution and factors influencing it. Staples distribution was often well above the subsistence level of around 17 kg of paddy per capita per month. The level of gross deliveries to the state was not simply related to the level of distribution to cooperators. This suggested that such advanced cooperatives could sometimes be favorably treated. It also supported the official contention that taxes "tended to be fixed for a period of years."[4] In these cooperatives the state had not immediately appropriated suprasubsistence levels of net production.

According to official statistics, the average proportion of gross paddy production supplied to the state for the DRV as a whole in the two years 1974–75 was 17 percent.[5] This is probably inaccurate because of the likely overestimates in DRV output. DRV data on procurement are likely to be somewhat less biased than the output statistics, suggesting that the DRV procurement percentage underestimates the true figure. The presence of on-the-spot observers should have reduced biases in the sample data. Thus, whereas the cooperatives of the sample seem to have delivered a comparatively large share of their production to the state, this conclusion is not very robust. There was very little information on any counterpart deliveries to them from the state although the bonus systems operating on collective labor depended in part upon the allocation of such goods.[6]

In principle cooperatives were not allowed to sell direct to the free market, and the sources did not mention such behavior. There was some indication of sales of artisanal products by the cooperative to cooperators; Dai Thang (Xuan Thuy, AWM) was selling eggs. Another related and important area was nonstaples deliveries to the state.[7] The limited information on this was as follows:

Dai Thang (Quynh Luu, ASM). "Merchandise production" was 305 dong per worker, that is, ca. 0.27 m dong or nearly 25 percent of the value of gross output. The value of rice and other taxes was nearly 0.13 m dong.

Van Tien (ASM). The cooperative sold 10 percent of its tile production and 15 percent of its lime production to the state.

Thang Long (ASM). "Merchandise production" was 375 dong per worker, that is, 0.61 m dong or 46 percent of the total value of gross output. This was made up of staples, pork, bricks, tiles, carpets, potatoes, and garlic.

Giao Tan (AWM). The cooperative was selling over 50 tonnes of jute annually to the state.

Dai Thang (Xuan Thuy, AWM). In 1975 the cooperative supplied 41 tonnes of pig meat (value at least 100,000 dong) and 216,000 eggs.

Distribution Systems

The distribution system directly reflected brigade independence. As discussed above, this could be at either brigade or cooperative level. Two cooperatives in the sample—Giao Tan and Hai Nam—still had brigade-level distribution, but the latter abolished it in 1976. The other cooperatives used the prescribed centralized system of norms and labor categorization that accompanied cooperative-level distribution.

There were a number of possible outlets for cooperative output net of ordinary production expenses and after deliveries to the state. Cooperatives were supposed to make allocations to the accumulation and other funds and for communal services such as schools and dispensaries. About 20 percent of collective distribution to cooperators was intended for the "priority families" as pensions and other income supplements. These included support for the sick or young, the families of war-dead or wounded soldiers, and, lastly, cooperative and commune cadres.[8] Two other important items supplemented the distribution of collective output as direct payment for collective work: (1) extra payments to cooperators in rice for the provision of such items as manure and pig meat from the household economy; and (2) payments in rice to the cooperative by the cooperator for the right to have extra access to collective assets, most importantly land.

Cooperatives' penalty-bonus systems varied greatly. In part this reflected the importance attached to the fulfillment of labor duties.

Brigade-level distribution. In this system the brigades essentially shared their net product among their members. Incomes could therefore vary substantially between brigades according to land yields and population. Individual incomes depended upon the extent of labor sharing. An extension of the system allowed the Management Committee to finance production in other branches by paying labor in workpoints negotiable at the brigade in which members of the worker's family worked. This was happening in Giao Tan:

> Giao Tan, *like a number of other cooperatives*, does not pay for subsidiary branch work via a common system of distribution throughout the cooperative but relies upon the cultivation brigades. This leads to an irrationality in that two people doing the same work do not earn the same. We know that between each cultivation brigade there are differences in output and in penalties and bonuses. Furthermore, in a number of brigades there is "endure the work and defend the points," leading to nonobservance of the norms and resulting differences in the value of the labor-day. Workers in the subsidiary branches do not particpate in cultivation, but still have to suffer those penalties and bonuses. In brigade no. 8 the labor-day was worth 1.6 kg of paddy, in brigade no. 12 only 1.3 kg. . . . This system of distribution clearly does not encourage the development of production and also reduces incentives and competition, even leading to contradictions between people working in the same line.[9]

This system had positive implications for the relative strength of the cooperators' position vis à vis the Management Committee in the collectivity. In this cooperative the committee was facing difficulties in controlling the assembly—cooperators would not approve the plans for financing development from the cooperative's accumulation fund. This was not a blanket opposition to such deductions from collective output: resources allocated to the social fund were spent on creches, schools, and the "club." But the likely benefits from the accumulation fund were not valued so highly. Cooperators would not agree to make contributions. So, "to guarantee a correct distribution and fully utilize its effects on production and living standards, on the one hand the cooperative must explain to its members so they understand the material consequences of setting up the undistributed funds and the social fund, and on the other *the cooperative must be firm and avoid not daring to resolve this rational and necessary problem because of a few objections.*"[10]

Management Committees also encountered the consequences of brigade-level distribution when they tried to use the brigades to control labor. Brigades tended to "want all their product back" and would not as a result economize on work points; the cooperative's penalty-bonus system "had little effect." The manager at times earned less than his deputy because Management Committee cadres were attached to brigades. Brigade incomes therefore determined the value of each cadre's workpoint allocation. This was felt to be deeply irrational. But despite the brigades' apparent power, the cooperative both organized and financed a system of supplementary incomes. This allowed it to sell staples to the needy. There was a fourfold categorization of food-deficient families:[11]

1. Families of war-dead or soldiers fighting "far away," who received 100 percent of their needs.

2. Families of key cadres, who received 90–100 percent of their needs.

3. Ordinary households who had ill members or many children, who received 80–90 percent of their needs.

4. "Lazy people" or those who did not accept the cooperative's management and worked elsewhere received an amount and a price that were "subject to assessment."[12]

The wider implications of Giao Tan's experiences are consistent with general arguments. Brigade-level distribution occurred when the balance between the brigades and the Management Committee favored the former. The brigade leadership's greater control over output then reinforced their position.

Hai Nam and Giao Tan illuminate the wider consequences of a high degree of brigade independence. Neither cooperative had fully reorganized its brigades' landholdings. There was an established "trade" in ploughing services between Hai Nam's brigades at a price of 1.5 dong and a bowl of rice per man per animal per day.[13] The relative ineffectiveness of Giao Tan's central accounting system was indicated by a recommendation to end the system where each brigade had a drying yard and a warehouse of its own and centralize these in one place "in order to be able to supervise closely the productivity of each brigade."[14] The Management Committee would not control "productivity" at the threshing yards and warehouses, but illegal expropriation.

Another incident that reveals the distributional effects of weak management concerned the III-point contract Giao Tan put out to its number 5 brigade for the 1974 Xth month harvest. The Management Committee had given out a contract for the expenditure of 10,000 marks, but the brigade only really worked 9,000. The remainder were "avoided," partly by economizing, but mainly by ploughing only once when they should have ploughed twice. The spare 1,000 marks were "freely divided among some members of the brigade."[15] This hints at the existence of alliances and divisions within the brigades. The source did not explain what these were, nor why only "some" members of the brigade benefited.

Cooperative-level distribution. The majority of cooperatives in the sample used the cooperative-level distribution system. Vital elements of it were the norms and labor categories that established: (1) the quantity and quality of labor required for formal recognition that a labor-day had been worked (a record to that effect could then be entered in the brigade's account book), and (2) the precise number of points paid for that piece of work, with any supplementary penalty or bonus.

Cooperatives frequently categorized individuals as well as work. This allowed them to match certain types of worker to jobs thought appropriate for them. There was some relationship between this and some cooperatives' practice of requiring different amounts of duty-labor (measured in terms of labor-days) from different groups of cooperators. A number of penalty-bonus systems of varying generosity and severity were in operation. These operated on groups as well as on individuals.

Norms. The existence on paper of a system of norms revealed little about their use in practice. Since they were a required part of the management apparatus, reports to higher levels looked better if it could be said that "X cooperative used Y norms." The number used also depended as much on the extent of subsidiary branch development as on the sophistication of the cooperative's labor management methods. It varied greatly, from 953 in Dai Thang (Quynh Luu) to 486 in Dai Thang (Xuan Thuy).

In Van Tien, where the Management Committee was well-established, the norms were "a product of experience." The cooperative had set them up over a period of six or seven years. It had deliberately adjusted them over time to help equalize the incomes of cooperators working in different branches, and it had increased norms appropriately. But in Hai Nam, where the Management Committee was far weaker, the norms were said to be "too easy"; the ploughing norm was for a six-hour day and had not been changed for seven to nine years. The position of the norms as a focus for conflict was also apparent in Vu Thang (M), where the Management Committee had "amplified" the system by taking greater account of particular working conditions and by reducing labor targets "as a bonus."

Reducing a norm had two main direct effects. It effectively increased real remuneration if distributed output and the workpoints allocated for fulfilling the norm were both unchanged. The cooperator received the same real income for less work. But if the cooperative paid out more workpoints in order to increase labor inputs and thereby compensate for the lower norm this would devalue the workpoint. The associated fall in income had then to be compared with the direct benefits of the lower norm when assessing the overall distributional impact. A second effect was to change the real burden of labor-duties. A lower norm increased the number of labor-days required to complete a particular task and reduced the effort required to complete a labor-day. Since labor-dutues were usually expressed in labor-days this allowed them to be done more easily.

Labor categorization supplemented norm adjustment as a means of altering labor remuneration. The main classifications of tasks and of laborer were as follows:

a. A task classification (*bac**), usually into six or seven grades. This was noted in Dai Thang (Xuan Thuy), Dai Thang (Quynh Luu), and Vu Thang. This was in addition to

b. A cooperator categorization that depended upon the "strength" and "skill" of the worker and therefore to some extent upon the categories of work done. High-value work was done by skilled laborers, but the cooperator categorization seemed often to be a way of paying more or less for other reasons, most especially sexual discrimination. Thus in Dai Thang (Quynh Luu) there were three categories of laborer, and men could not do work that paid less than some 15 percent above the minimum paid to women.[16]

Many cooperatives used some system of penalties and bonuses coupled with ad hoc adjustments to points (e.g., during bad weather) to determine points actually paid to a particular worker for a specific job. Such adjustments were the usually the brigade's responsibility.

The penalty-bonus systems. Penalty-bonus schemes were quite important parts of a cooperative's battery of measures designed to encourage collective labor. So far as any individual worker was concerned, these acted in two quite separate areas. These were, first, the penalties and bonuses aimed at making him or her actually work, and, second, those designed to make him or her work well. The former were comparatively rare and were mentioned in the sample only in the highly disciplined Dai Thang (Quynh Luu), where the basic period of work was ten days:

> Workers participating adequately for ten days are granted an additional 2 dong, but if they do not do so they are penalized immediately after the ten-day period finishes. At harvest time a worker who is absent without reason for one day is fined 6 dong, while there is a "nurturing" of workers during that period of 5 hao a day. This leads labor to participate in work with discipline and organization, and with this pressure on, there is a high productivity.[17]

This was not the only device used to encourage the fulfillment of labor-duties. The cooperative was comparatively well-off, with an average paddy distribution of 22.5 kg per capita per month. A system of food rationing discriminated against workers who preferred to work on their own account rather than for the collective. Those who worked 60–80 percent of duty-days received 100 percent of rations; 80–100 percent of duty-days, 120 percent of rations; 100–120 percent of duty-days, 150 percent of rations; over 120 percent of duty-days, 170 percent of rations. Cooperators who worked less than 60 percent of the labor-duties received no staples rations, although the old, young, and weak did obtain access.[18]

Most cooperatives in the sample did not operate such a system. Penalty-bonus schemes usually concentrated upon trying to make cooperators work better when they had actually turned up for collective labor. They applied directly to the work done and were far less severe. Management Committees tended to be unable to operate them unless relations with the brigades permitted it. As expected, the intensity of penalties and bonuses and the extent to which they were actually applied tended to vary with the degree of dominance of the brigades by the committee and the resulting orientation of the brigade leadership toward higher rather than lower levels.

Labor Shedding

For some cooperatives of the sample, data on collective labor mobilization were comparatively abundant. These can be used to illustrate changes in labor use in these "advanced" cooperatives. Unfortunately, however, the data themselves have limitations, and many cooperatives' accounting systems exhibited considerable peculiarities. Any conclusions must be extremely tentative.

Rice-producing brigades in nominal cooperatives could be strikingly inefficient labor users when judged according to the labor requirements implied by the specified norms. This was partly because of their emphasis upon labor sharing to ensure subsistence entitlement. An additional factor was the lack of strict supervision by brigade leaders whose economic sources of authority, in nominal cooperatives, did not depend upon a close control of labor. In fact the position was quite the reverse, with minimal controls in collective labor facilitating cooperators' own-account activities.

In active cooperatives a new and modernizing "technocratic" group might wish to supervise labor more closely in order to monitor the introduction of new production methods (see chapter 3). In any case, a newly active cooperative would almost certainly need sharply higher effective labor inputs as the collective economy expanded. Strategies of self-reliance would result in a need for large-scale labor investment in direct accumulation. The resulting increased demand for labor in the collective sector could, however, be offset by a greater labor effort from a reduced nominal labor input, whether in terms of labor or in labor-days. This might derive from reduced "labor sharing" and would probably be accompanied by changes in the pattern of material incentives. It is therefore extremely interesting to observe declines in some measures of collective labor.

In some cooperatives, reported labor mobilization levels fell rapidly. Well-documented examples are the two Dai Thangs and the model cooperative, Dinh Cong. All three had well-developed management structures, but in Dai Thang (Xuan Thuy) managerial control was weak. Production diversification in all had

Table 7.4

Reported Labor Use in Cultivation, Dinh Cong

Year	Paddy production (tonnes)	Days worked per tonne	Implied total	Permanent workers in cultivation	Implied days per worker[a]
1970	1,215	160	194,000	790	244
1972	1,490	150	223,500	771	290
1975	2,060	146	300,470	527	390

Source: Le Trong (1978).

[a]Since the figure for the total number of workers included those producing the ca. 30 percent of output in the branch that was not rice, the true figure for the number of labor-days worked p.a. was higher than that calculated.

reached the stage where the cultivation branch produced less than 75 percent of gross accounting output. Table 7.1 shows that the numbers reported working for the cooperative as a percentage of those understood to be subject to labor-duties had gone down to nearly 60 percent in Dinh Cong and Dai Thang (Quynh Luu) and was below 70 percent in Dai Thang (Xuan Thuy).

Thus, even in these advanced cooperatives approximately one-third of the available labor force apparently did not register for collective work. But these statistics do not accurately measure labor mobilization because of the practice of paying "nonpermanent" labor in workpoints allotted to workers in the same family. The reported "labor supply" (*suc lao dong*) of such cooperatives, which probably corresponds to the DRV statistics for "cooperative labor," was basically the working-age members of families that had joined the cooperative. Cooperatives recorded labor use, however, either in terms of the numbers who both "submitted permanently to the cooperative's management" and were in direct receipt of workpoints for work done, or through the number of workpoints awarded in a branch for work done. If peak-period labor or subsidiary-branch labor was paid for in workpoints negotiable by a member of the same family, it could be possible for cooperators to appear to work a number of days far greater than that humanly possible. This can be seen from the example of Dinh Cong's cultivation branch (see table 7.4), where "permanent" workers in cultivation were paid for 250 days' work in 1970 and then 390 in 1975. The number of such workers had, however, fallen sharply over the same period.

In Dai Thang (Quynh Luu) circumstances were slightly different. While the numbers in permanent employment in the staples branch had not fallen, evidence for the use of nonpermanent labor comes from the workpoint data. Each worker in the cultivation branch earned 344 marks in 1976, which implies a total of 0.17 m marks. But the actual production needs were reportedly 0.30 m marks. Clearly, the cooperative had directly credited extra labor with workpoints.

Table 7.5

Permanent Labor Force by Branch, Van Tien

	1973	1974	1975
Cultivation	615	613	583
Livestock	119	30	36
Subsidiary branches	1,311	1,442	1,549
Total (excluding admin. etc.)	2,045	2,085	2,158

Source: Tran Dinh Thien (1976, 69).

In Van Tien the decline in "permanent" labor in the cultivation branch continued despite an already extreme concentration of labor into the subsidiary branches (see table 7.5). The cooperative was benefiting from prerevolutionary experience with artisanal production (see chapter 9).

The evidence from other cooperatives was patchy. Vu Thang, the other model cooperative, had increased the number of labor-days worked per worker from 332 to 346 between 1970 and 1977; in the early 1960s it had been as low as 180.[19] Almost all of these cooperatives, or at least their smaller predecessors, had been rice monoculturalists in the early 1960s. Even allowing for the use of nonpermanent labor at periods of peak labor demand, the comparatively low proportions of the workforce permanently attached to the cultivation branches is striking (see also table 7.1). It reveals the extent of the difficulties involved in solving the surplus-labor problem by creating additional employment opportunities in the collective sector.

The available information confirms that substantial changes in labor utilization often occurred during periods of production diversification and rising yields. The number of cooperators recorded as permanently working for the cooperative tended to stabilize or even decline. The number of days attributed to them rose. The latter apparently resulted from the combination of an increase in the number of days work actually done by them, and the use of supplementary labor either directly credited with points or paid via permanently employed family members. Although the precise balance between these two strategies is unknown, the result was often strikingly low recorded participation rates, as measured by the proportion of the total population actually registered as being employed by the cooperative. This varied from 19 percent in Dinh Cong (excluding some minor subsidiary branches) to 41 percent in Van Tien.

This has wider significance for the local power balance and its interaction with patterns of economic development in such cooperatives. One possible explanation would accompany the view that the importance to the Management Committee of a high level of social authority encouraged it to seek better direct control of labor by increased supervision and regulation. This would be more important in cooperatives with powerful and orthodox Management Committees. From this it would follow that the need to control production and enforce technical norms tended to drive the management team into relying upon a relatively small number of workers. Use of disciplinary norms may have meant that cadres preferred a smaller number of efficient workers who worked better to a larger number who would each have to be given an opportunity to earn their points and so could not be so effectively supervised. If this interpretation is correct, the reported use of nonpermanent labor at peak periods would be in the more easily supervised tasks. Complaints of "surplus" labor would be more prominent in tightly managed cooperatives. Growth in employment opportunities accompanying production would obscure the latter point. It would be noticeable, however, that as the power of the Management Committee grew, so its perception of "underutilized" or "superfluous labor" (*lao dong roi rao**) would increase.

An alternative explanation would point to the possibility of restriction of entry to the cultivation brigades by the permanent labor force. This would be more important in cooperatives where cooperator interests had a loud voice. As workers in the branch moved out of it into the subsidiary branches, or left for other reasons, it could have been in the interests of those who remained to take on as replacements only nonpermanent substitutes if this led to a redistribution of incomes in their favor. In the absence of systematic intrafamily rivalries, the payment of such nonpermanent labor via the permanent family member could largely offset such redistributional effects. The cooperative's management team could also use lower rates of remuneration to nonpermanent labor as a means of reducing incomes from collective cultivation and thus freeing resorces for use elsewhere. The conditions required for labor to be recorded as permanent remain unknown in any detail, but changes in them could have provided one means of reducing labor remuneration by shifting to "temporary" labor. It could also be possible, however, that greater competition with own-account activities for such "mobile" labor could at times raise rates of pay for such work above that awarded for those permanently employed.

The pattern of development of production throughout the collective was arguably of considerable importance. If cooperators were experiencing rapid income growth in the subsidiary branch, then this could reduce incentives to belong to the permanent labor force in the cultivation branch. In addition, there could also be a fall in the labor requirements there because of increased use of capital financed by the subsidiary branches. Because of widespread shortages within the state sector the latter was apparently rare outside model cooperatives. Furthermore, a centralized distribution system in principle ensured that the value of the work point in all branches was approximately similar. In practice this was often not the case. Factors determining labor participation must have included the wider context—the ability to dodge labor duties, "privatization" of collective assets, and so forth. Labor incentives encouraging movement into the subsidiary

Table 7.6

Labor-days Recorded per Tonne, Paddy

Cooperative	Labor-days per tonne
Dinh Cong (M)	146
Dai Thang (Quynh Luu, ASM)	194[a]
Van Tien (ASM)	180[a]
Hai Nam[b] (AWM)	95[a]

Sources: See bibliography, table B1.

Note: All data for 1975 except possibly Hai Nam (date unspecified).

[a]Labor-days recorded in terms of marks, equivalent to ten workpoints.

[b]The data for Hai Nam are unreliable because there was no cooperative-wide distribution system.

branches were most likely also highly dependent upon working conditions, such as cleanliness, lighter work, and ability to wear nonfarming clothes. These could provide powerful additional reasons for labor to leave the cultivation branch.

Conclusions

The level of collective labor mobilization was an important indicator of the overall position of the cooperative's management team. Labor was, after all, a basic economic resorce. The extent of the problems involved in generating employment in the collective sector are emphasized by the extremely low technical requirements of collective rice production. Table 7.6 summarizes the available information on the recorded labor-days per tonne of paddy.

Basing calculations upon the first three cooperatives mentioned in the table and ignoring intermittent and seasonal periods of peak labor demand, a cooperative with a population of 5,000 would require a labor input of 0.15–0.20 m labor-days to provide 15 kg per capita per month. This amounted to only about 150 labor-days per family. From this it is clear that rice cultivation, especially with the likelihood of increased labor productivity as a result of technical advance, would be incapable of providing adequate collective employment.

The major hope for labor absorption and employment generation was the subsidiary branches. Chapter 9 considers their development. An alternative strategy involved precocious development of nonstaples cultivation (as in Thang Long). Here, however, output expansion could depend upon the use of imported machines to overcome timing problems arising from the interaction between staples and nonstaples cultivation. During periods of peak labor demand, which could be shortened by changes in cropping patterns, there were limits to the extent to which higher labor inputs could cope with bottlenecks: "If there are no machines, the cooperative will find it hard to be able to sow [as much as] 70 percent with early Xth month rice, and because of this it will be hard to use [as much as] 70 percent of the land for a winter crop.''[20]

Supplying the machines needed to overcome such limitations on the expansion of nonstaples cultivation would partly offset the employment-creating effects of output expansion. The final outcome would depend, however, upon the overall consequences for the cooperative of such a strategy (for instance, the ability to buy-in raw materials for the subsidiary branches conferred by increased sales from the cooperative).

The problems faced by any policy makers when confronting such an agricultural system would be extremely difficult to resolve. The day-to-day operation of the collectives further aggravated difficulties. The very low labor requirements in the dominant branch of collective activity accompanied systemic labor-motivation problems. The effects of the strength or weakness of the management team in its relations with the brigades and the cooperators were of crucial importance in overcoming the frequent unwillingness of cooperators to participate effectively in collective labor.

8

Pig-Rearing as a Focus for Conflict

Pig-rearing was an extremely important part of cooperator own-account activity. It was a major source of the cash needed for such vital expenses as weddings and other life-events. At the same time, official policy clearly and firmly asserted the importance of collective pig herds. These were intended to provide a basis for increasing the cooperative's mobilized surplus in the form of pig sales to the state at the prevailing low state prices. Although policy asserted that these would be exported, in practice they were usually eaten before they reached the harbor.

The Planned Pattern of Development and the Economics of Pig-Rearing

Experiences within the sample of attempts to increase production of pig meat and manure help to illustrate underlying behavioral patterns (see table 8.1). Ideally, a cooperative's livestock branch had a strongly integrating function. A centrally located pig herd should have allowed technical cadres to supervise the introduction of new techniques—foreign strains, better care of piglets, artificial insemination, systematic records of the effects of different feeds, and so forth. Manure was to be collected and mixed with vegetable matter to create fertilizer for use on the fields—an additional input for the rice brigades. The Management Committee had to organize feed supplies for the cooperative's pigs—paddy, vegetables, or waste. The committee also had to find supplies of other means of production (especially piglets). Here it usually had to choose between the cooperative, the household sector, or some outside source.

In practice, local views of the likely consequences of possible alternatives largely determined the outcome. These calculations were often hidden from view. Cooperatives' accounting methods tended to obscure the real costs and benefits involved in different strategies. Cooperatives normally had to use official prices in their internal accounting, and these accounting prices were far below free-market prices. This was consistent with the NMS's overall attitude toward the "price problem." (The distributional effects of this will be discussed shortly.) In addition, however, in a highly politicized economy the low "shadow" price used to compute, for example, the benefits of extra pork production frequently ignored important side-effects. Policy implementation could result in improved access to scarce state goods. A supply of pork could allow local cadres to acquire influence with superior levels through gifts or feasts. It is hard to quantify such factors, but the low official state duty price used in cooperative accounts to calculate the gross benefits of production ignored them entirely.

Integration of collective pig-rearing into production elsewhere in the cooperative would reduce brigade autonomy. The rice brigades would then depend upon outside supplies of means of production. Brigades had an interest in resisting this trend. One result was that they might allow resources such as manure to go to waste rather than permit them to become a means for enhancing the Management Committee's authority. Here the prices charged for such inputs within the cooperative would be important.

When brigades were resisting establishment and development of the collective pig herd, a Management Committee had a number of alternative strategies. These essentially involved either vertical integration of pig-rearing within the collective economy or use of the household sector.

Vertical integration was one response to difficulties in securing resources from areas outside a Management Committee's direct control. Such a strategy basically aimed to extend the committee's direct authority over a wider area of the pig-production process. It was therefore often another example of the use of administrative measures to overcome adverse material incentives. The committee would usually allocate land directly to the specialized brigade responsible for the pigs. The brigade would use this land to grow feed, and in practice it would also use on it most of the manure produced by the pigs. Like the rice brigades, the specialized brigade teneded to seek self-sufficiency. This would be contrary to the overall intentions of official policy, which sought a strongly integrating role for such production diversification. But since it did not directly threaten the autonomy of the rice-producing brigades, it might be more acceptable, although it relied on inducing somebody to give up their land for use by the specialized brigade. In such circumstances the absence of forms of exchange (of manure, for example) could reduce the influence of incorrectly set "shadow" prices on resource allocation decisions. Apparently unprofitable pig herds could more

Table 8.1

Pig-Rearing as a Focus for Conflict: Summary of Main Findings

Issue	Time-series	Cross-section
Collective pig-rearing	Collective herds often established but frequently encountering difficulties	Severe problems in weakly managed cooperatives. Spot observations suggest pig-rearing often of low or negative profitability (but accounts unlikely to include all relevant costs and benefits)
	—	Strength of management affects (1) ability of cooperative to overcome unfavorable material incentives; (2) ability to ensure efficient use of resources mobilized into collective pig herd
Private pig-rearing	Growth of supplies from private herds to collective sector in some cooperatives	Need to use free-market paddy-prices to overcome adverse incentives when official prices used
Response to difficulties to developing collective herd	Integration within collective branch. Use of developing household sector and free-market herd: incentive levels	

easily persist, perhaps allowing advantages accruing outside the accounting system to continue.

Cooperators' statutory right of free-market disposal apparently played a major role in the second strategy alternative, which was based upon use of the household economy. Cooperatives were denied this right.[1] Management Committees who had some control over output distribution could use this to improve the material incentives acting on cooperators to supply resources to the collective. This could be done by exchanging paddy for household products. In surplus-producing cooperatives the Management Committee could enter such outlays in the cooperative's accounts at state duty prices but expect that the cooperators would value them at the far higher free-market prices. For this and other reasons, reliance on the private sector probably tended in many instances to appear more profitable (in an accounting sense). Labor inputs may also have seemed cheaper, given the widespread problems encountered in mobilizing labor into the collective sector. At the same time the pigs' feed, being household waste, would have had a low opportunity cost within the collective economy.

The precise levels of the shadow prices used in a cooperative's accounts played a crucial role.[2] During the period under study the duty price of pork was 2.4 dong per kg live-weight; the duty price of rice was 0.32 dong per kg (the incentive price was 0.90 dong per kg). A fundamental question was the profitability of collective pig herds fed on rice—a phenomenon observable in some cooperatives, and especially in the model cooperative Vu Thang. A pig was expected to eat 0.8 kg of rice feed per day (rice or rice bran).[3] A herd of 250 could provide around 15 tonnes of meat annually,[4] yielding 36,000 dong for an input of 73 tonnes of feed that would cost 23,500 dong if it were all paddy valued at the duty price. This gave a net income of 12,500 dong. At a free-market price of 5,000 dong a tonne, this would buy 2.5 tonnes, which would feed around ten people for a year—not a very large real return. But there were as yet unincluded costs, including labor, secondary feed, and buildings, offset by benefits from the fertilizer.

The sources do not permit a proper investigation of the other costs of collective pig-rearing. They were not likely to be small. The conventional wisdom was that a herd had to have at least 1,000 pigs before it could be profitable, even using established advanced techniques.[5] It is possible, however, to examine the accounting benefits from the resulting manure.

A pig was expected to yield around 4 tonnes of manure annually.[6] This was meant to be valued at around 10 dong a tonne but was sometimes valued at a higher price—in one case, 18 dong.[7] Thus the herd of 250 could yield an additional imputed benefit of 10–20,000 dong from its manure, implying that the accounting price of manure was of considerable importance in the profitability of collective pigs. It is surprising that the accounting values of pork and manure are of such similar orders of magnitude (36,000 dong compared with 10–20,000 dong). This had interesting distributional consequences. The rice brigades had in effect to pay the cooperative a high price for manure from the pig-rearing brigade. By setting internal prices in this way, and keeping rice prices low, the Management Committee's capacity to prevent the brigades from retaining output was enhanced. It therefore helped to increase centralized control over resources.

Experiences in the Sample

Many cooperatives were experiencing severe problems in developing collective pig production. Active cooperatives tended to use the household sector more often than other cooperative types and frequently viewed the collective herd as a marginal activity, if not a positive burden. Collective pig herds were almost unknown in nominal cooperatives but, reflecting official policy, were well-established in model cooperatives (see table 8.2).

The sources focused upon two main factors constraining development of the collective herd: the supply of piglets and the supply of feed. In addition, the supply of fertilizer from the pigs was rarely as great as it should have been. The tendency to use the limited manure supplies on land set aside to grow feed-crops for the pigs (if it existed) tended to reduce their availability for rice cultivation.

The first constraint was visible in Dai Thang (Xuan Thuy), Van Tien, Hai Nam, and Giao Tan. In Hai Nam difficulties in securing supplies of piglets were attributed directly to weak Management Committee control. The two sties from

Table 8.2

Pig-Rearing in the Livestock Branch, ca. 1975

Status	Vu Thang (Model)	Dinh Cong (Model)	Van Tien (ASM)	Hai Nam (AWM)	Giao Tan (AWM)	Dai Thang (Xuan Thuy) (AWM)
Size of household pig herd	1,670	n/a	1,960	1,700	n/a	1,800
Pigs per family	1.7	n/a	1.9	1.5	n/a	0.94
Household sales of pig meat to cooperative	n/a	n/a	90%[a]	n/a	n/a	11 t
Household sector sales of piglets to cooperative	n/a	n/a	Yes	No[b]	Yes	16 t
Household sector sales of manure to cooperative	4,000 t	n/a	4–4,500 t	3,400	n/a	6,136 m³
Sales per pig	2,400 kg[c]	3,410	2,200 kg	2,000 kg	n/a	3,410 kg
Collective herd	1,480	1,206	350	500	120	250
No. of sties	n/a	n/a	1	2	1	1
Condition of sties	n/a	n/a	n/a	fair	sub-standard	poor
No. of workers	<137	n/a	<36	35	20	7.33
Pigs/worker	>11	n/a	>10	14	6	7.8
Pig meat output	n/a[d]	n/a	200 t[e]	n/a	7,600[d]	14 t
Manure output	5,760 t	n/a	250–300 t	40 t	n/a	112 m³
Output per pig	3,900 kg	n/a	780 kg	80 kg	n/a	448 kg
Sows in collective	n/a	Yes	14	34	n/a	none
Source of piglets	n/a	n/a	60% from households	"outside"	house-holds	house-holds

Source: See bibliography, table B1.

[a] Of total output. The source gives total production at 4,000 tonnes (*sic*), of which 400 came from the collective sector.
[b] The two sties of the cooperative operated independently; one sold piglets outside the cooperative while the other had to "buy in." The source of these supplies was not given.
[c] The duty was 3.6 tonnes per worker in single-worker family units, falling to 3 tonnes per worker in three-worker families.
[d] Equivalent to 3.2 tonnes at the duty-price of 2.4 dong/kg (Hanoi I).
[e] Total production in other sectors was 107 tonnes.

the separate cooperatives from which Hai Nam had been formed in 1973 were functioning separately and independently. In none of these four cooperatives had the 1,000 level remotely been reached. Van Tien, Giao Tan, and Hai Nam were meeting direct losses, while in Dai Thang the situation was far from satisfactory.

The problem of securing adequate supplies of piglets of a sufficient quality was of particular interest. In practice the Management Committee had only two real alternatives—the household sector or a boar and sows in the collective herd.[8] There were clear conflicts. Complaints about the household sector's shortcomings as a source of piglets resulted in pressure to expand the collective herd. This was another instance of tension in the relations between Management Committee and households, leading, as expected, to pressures for a vertical integration of collective activity. But in additon to the Management Committee's inability to enforce priority supply to the collective herd, the management of the herd was often criticized. Dai Thang (Xuan Thuy) provided an interesting example. On the face of things, the households were a far better source of pig meat and manure than the collective herd, but

> The important problem is the amount of meat supplied to the state, of manure supplied to cultivation, and piglets. Collective rearing must be the major source of meat supplied to society and of manure to cultivation. In reality the situation in Dai Thang is just the opposite. . . . This situation is quite contrary to the law of development of livestock rearing to large-scale socialist production.[9]

Such arguments simply gloss official policy rather than providing a proper justification for collective pig-rearing. To some extent such authors resolved their difficulties by taking the position that the collective herd must "lead" (which allowed for coexistence) and be an example. This line of thinking then led on to the view that the need to introduce new strains tended to imply reliance upon the collective herd because supplies of new breeding stock were state-controlled. Therefore the socialist sector was likely to have at least one unassailable basis in its monopoly control over these valuable new inputs. Greater reliance on the household sector was, therefore, less of an ideological threat.

The piglets supplied by the households in Dai Thang had problems adapting to life in the collective pens. The percentage of runts was also very high, averaging around 20–25 percent. The percentage rose to 33 percent in 1975, a bad harvest year. In fact the cooperators never supplied enough piglets, so that the Management Committee could not realize the cooperative's plan: "Thus, with regard to both economy and the organization of production, the production of piglets in the household sector for supply to the collective cannot meet the requirements of large-scale production in livestock."[10]

The official commitment to an expanded and integrated production system was strong. It was therefore natural to justify any production integration in accordance with the principles of the NMS in terms of the inherent value of increased production of means of production within the collective.[11] But such arguments ignored deep-rooted underlying issues: there was no desire on the part of the relevant households to relinquish control over the supply of piglets. Note the role played by some of the members of the Management Team, in this case the cadres responsible for the collective herd: "Inheriting the piglet-rearing tradition of the region, the pig-sty of the cooperatives has three middle-level technical cadres, with the ability to improve the collective stock by selective cross-breeding. But, in reality, in past years the technical cadres of the station '*have had severe shortcomings.*'"[12]

Other cooperatives were also experiencing difficulties with the households' supplies of piglets. In Van Tien a major problem was that the old strains were eating too much, and this exacerbated the feed shortage. The cooperative already had a boar and fourteen sows,[13] unlike Hai Nam, where one of its two stations was busy selling piglets outside the cooperative while the other remained totally dependent on outside supplies. The plan was to expand the latter so as to have 260–270 high-quality sows (from the existing core of 34), and to supply piglets to both the cooperative and the households. The small herd in Giao Tan was also relying on poor-quality piglets supplied by the households.

Questions surrounding piglet supplies thus had a number of wider implications. The cooperators, as expected, were often opposed to the Management Committee's attempts to control this particular means of production. The collective pig herd was indeed a way of introducing new strains, both to replace the households as suppliers of piglets and to raise the quality of the household herd. But there was little evidence that the latter had been happening. All in all, it is clear that this was not an area where the will of the party dominated, even in these advanced cooperatives; the households often did not supply enough piglets, and local interests were clearly dominant. This partly reflects the importance that cooperators attached to this key source of cash incomes.

The question of feedstuffs was more complicated. There were four possible sources: (1) a share of the rice crop; (2) vegetables from outside the livestock branch; (3) vegatables from inside the livestock branch; and (4) various waste products, especially bran from rice milling.

The initial steps in developing the collective herd relied upon 1 and 4, the first of which was in practice ridiculously costly, unless the cooperative based its calculations upon accounting prices that did not reflect the real costs and benefits involved. This resulted in pressure for an expansion of 2 and 3, often using manure from the collective herd to grow vegetables on land set aside for the purpose. A widely unimplemented decree of the early 1970s set aside 15 percent of the rice land in all cooperatives for vegetable-growing intended for the collective herd. There was also probably a directive on the proportion of paddy production that was to be set aside for pig feed.[14]

In Dai Thang the supply of feedstuffs was not secure. The availability of land was not an issue. The cooperative used only 7.4 mau (around 2.7 hectares) to grow feed for the pigs although around 10 mau had been set aside for that purpose.[15] There were feed shortages in November-March, when the pigs had to eat seaweed and water-potatoes. But generally, "Each month the livestock [branch] had to fetch paddy from the cooperative's warehouse according to the needs of the pigs for that month." Supplies were not guaranteed, and the collective pigs were the first to go when there was a harvest failure:[16] "When the harvest is lost and the volume of rice available falls, then so does that available for livestock, or if for some reason the cooperative needs staples, then the amount allocated to livestock is mobilized. Thus the source of feed for livestock is terribly unstable.''

In 1975 the rice yield fell from 5.2 to 4.1 tonnes per hectare and the collective pig herd suffered badly. Numbers declined from 400 to 250: ''In late 1975 there was bad weather. . . the group in chage of . . . rice seedlings was incompetent, the seedlings were soaked three times but did not sprout, so the paddy for the pigs had to be taken to be used as seeds.''

In this cooperative a team within the livestock branch grew vegetables. The livestock brigade was divided as follows: five workers caring for pigs, eight workers growing green vegetables, one worker cooking vegetables, and three middle-level technical cadres, for a total of twenty workers (including n.i.e.). The brigade also used bran from a rice-milling contract with the state that used a rice mill owned by the cooperative.

Other cooperatives faced similar problems with feed supplies. Van Tien was supplying 60 tonnes of paddy annually (probably of poor quality) to the pigs— ''this could not be increased''—and so they were turning to an increase in winter crops, that is, potatoes and maize. In Hai Nam there had been little effort to supply vegetables to the collective pig herd and the winter crop had not been developed. Giao Tan was not even using the existing resources properly: according to one source (Tran Quang Lam 1976, 15), the potential was for 600–700 animals: ''20 percent of the paddy milled for the state *plus* 5 percent of paddy production *plus* 100 hectares of maize *plus* 3 mau of vegetables . . . would be quite sufficient.'' But at the time of writing there were only 120 pigs; the twenty people looking afer them were ''three times what was necessary.''

While the intention was to shift away from paddy as a feed, the outcome was frequently the contrary. In Hai Nam the collective pigs each received just over 100 kg of paddy annually, while in Vu Thang they were getting 250 kg. The prime feed requirements varied with the composition of the herd, but figures of 0.8–1.0 kg a day were common. This corresponded to ca. 300–350 kg annually; Vu Thang was in effect using its high rice productivity to finance the collective pig herd from within the collective economy.

The problems associated with feed and piglets were fundamental, but it was obvious from the wide variation in labor use even in this small sample that there were also large differences in labor productivity. This reflected the attitude of each Management Committee—the main source of pressure to enforce norms and labor discipline in the collective economy. Pham Bich Hanh (1976: 20) was again revealing about Dai Thang: ''Principally because of the unstable supply of feed, the style is unhygienic; the manure cannot escape and makes the pigs undersized while wasting a lot of feed; the costs per kg of meat increase.''

''Managerial incompetence'' was a first explanation given for the enormous variation in the volume of manure supplied by the pigs. Another was that the statistics were padded by the presence of waste in the manure, which would have helped the cooperators to fulfill quantity norms. Urine was simply allowed to run out of the sty.[18] Such appalling waste of resources in a food-deficit region can be explained by the high real costs of using them. The threat to brigade independence posed by production integration at the level of the cooperative provides a possible explanation for such behavior. It was not that the urine had no value, but that using it within the prescribed system would have encouraged unacceptable changes in the local power balance. As in many other instances, centralization of economic power in the hands of a cooperative's Management Committee involved taking that power from people whose interests frequently encouraged them to resist.

In light of the many difficulties encountered in attempting to develop the collective sector, it would seem natural to examine the possibilities of the household sector. At the DRV level this was the area where most development took place. In practice an active cooperative could obtain large supplies of fertilizer and pig meat from the household sector. This was one of the most striking characteristics of the more successful cooperatives such as Van Tien.

Private Pig-Rearing

Small-scale rearing of pigs by peasant households occurred before collectivization. A supply of pork was valuable both for sale and for consumption, especially in the feasts accompanying marriages, deaths, and other social occasions. Given the existence of substantial economic activity within the households of the collectivity, one of the ways in which the mobilization of a surplus via the cooperative could be effected was by an encroachment upon the household sector. Households were often subject to duties in terms of the provision of pig meat, piglets, and manure. In some cases Management Committees apparently paid well for these supplies, often by low-price sales of paddy. In general, reliance on the household sector was a more successful strategy than the establishment of collective pig herds. This last judgment is particularly true of those advanced cooperatives that were not model cooperatives, where collective pig herds clearly failed. A mixed approach was also possible, and Vu Thang was mobilizing its households quite extensively beside the thriving collective herd.

The household sector was a potential source of manure, pig meat, and piglets. In all of these areas households faced multiple alternatives. Free-market exchange opportunities existed, and they competed with supplies to the cooperative (whether the brigade or the Management Committee) and own-use. In Dai Thang (Xuan Thuy) households sold their best pig manure to other households. As a result, they fulfilled their quantity norms for supply to the cooperative at a very low quality—the manure was mixed with waste.[19] The payment systems used confirmed the expected existence of severe price problems. In Vu Thang 70 tonnes of paddy were being distributed yearly to families that surpassed their manure duties. Overfulfillment generated an extra 2 dong per tonne up to 50 percent, and then an extra 3 dong. In 1977 the household sector contributed 4,000 tonnes of manure. A rough calculation using an accounting price of 10 dong per tonne for the manure and the duty price for the paddy implies that 60 percent of the accounting value of the manure was paid out in the form of paddy bonuses. The free-market price of paddy was far higher, but there was no information on the free-market value of manure. Even in this model cooperative with high material incentives the household sector was only meeting 80 percent of its manure duty.[20]

The payment and duty systems for pork were better documented. Dai Thang (Quynh Luu) posted a duty for supply of both manure and meat. The households were allowed to use 50 percent of the 5 percent land as they wished but had to use the rest for a variety of purposes, including pig-rearing.[21] Each year they had to supply 2.5 pigs per family.[22] For each kg of meat sold to the cooperative each family was allowed to buy 2 kg of paddy, while in return for sales above duty the Management Committee gave them privileged access to consumer goods.

In Van Tien arrangements were complicated by the use of pig-meat duties as a form of rent on excess holdings of 5 percent land. Here the Management Committee posted duties for individual cooperators rather than their families. Each individual had to sell 15 kg of meat yearly to the state. For each 10 kg of meat over this duty, the seller received the right to buy 20 kg of paddy and an additional bonus of 2 dong. Dodging of the duty was penalized, and a shortfall of 10 kg led to a fine of 2 dong and the loss of 10 kg of paddy from the collective distribution.[23] Here an unorthodox but secure Management Committee can be seen using its tolerance of privatization of cooperative assets to obtain resources and help mobilize a surplus for sale to the state. This did not appear to weaken the committee's position.

The main effect of these incentive schemes probably acted via the paddy allowance rather than the financial reward. This conclusion is supported by the use of preferential access to consumer goods as a supplementary incentive. The paddy allowance persisted in cooperatives such as Vu Thang and Van Tien where the staples distribution to cooperators was comparatively high—20 and 17 kg per capita per month respectively. This leads to two important conclusions. First, Management Committees intended that cooperators sell such paddy onto the free market. They could then take advantage of high free-market prices in a way that the cooperative could not. The committees therefore overtly used the dual pricing characteristic of the aggravated shortage economy to attract resources into the collective sector (see chapter 1).

Second, there is here a key interaction between changes in the pattern of surplus extraction and changes in production conditions, in the sense of changes in the composition of output. The Management Committee's control over distribution allowances allowed it to regulate the production of the household sector.

Conclusions

Pig-rearing for sale to the state was widely seen as unprofitable at duty prices in both the collective and the household sectors unless cooperatives could build up very large collective herds with their own food supplies or use a supplementary distribution of collective resources to increase material incentives. It is probable that the apparent profitability of collective herds in some instances resulted from incorrect pricing of the alternatives (i.e., using the low state duty prices for rice) or simply ignoring the full opportunity costs of the resources committed. Collective herds could also appear socially profitable if control over pork was of particular importance and of more value to the Management Committee than paddy. Note that incentive sales of paddy to the state could simply lead to unusable credit balances while sales of pork—a strongly prescribed policy aim— may have had better results because of possible privileges to cooperatives successful in this area. Pork could also be used to cement alliances with higher levels in the rural administration (a frequent ''loss'' from the state distribution system), with other groups in the commune, or as an incentive to cooperators. Rising per capita incomes could also lead to a shift in demand away from staples to meat.

Pig-rearing for sale to the cooperative was apparently profitable for households if the cooperative paid them in paddy. Such behavior effectively used the right of cooperators to sell on the free market at substantially higher prices to subsidize the state's purchases of pig meat from the cooperative.

9

The Nonagricultural Subsidiary Branches— Collectivization or Taxation?

Issues

Management Committees' attitudes toward the various subsidiary branch activities[1] provide opportunities to observe the different implications of control over output distribution and over production (see table 9.1). In the weaker cooperatives little attempt was made to do more than license such activity. Prescribed policy, on the contrary, required the cooperative's management to involve itself in the detailed control of production. Highly orthodox Management Committees attempted this, often without great economic success. In intermediate situations a committee could use the cooperative's control over distribution (of consumer goods, means of production, and marketing) to encourage the development of subsidiary branch activity. Unfortunately, little is known about the interactions between this and other branches.

The various activities grouped in the branch divide conveniently into three groups: (1) Relatively independent activities that required no inputs from other areas of the cooperative but relied heavily upon outside supplies. Because of the great demand for building materials these were primarily bricks, tiles, and lime.

Table 9.1

The Subsidiary Branches: Summary of Main Findings

Issue	Time-series	Cross-section
Collectivization	Often related to wider issues in collectivization of commune	More strongly managed cooperatives tended to concentrate on reduced number of lines
Development of output and deliveries to state	Rapid development in some cooperatives, constrained by material incentives and raw material shortages	Highly variable
Centralization of production	Data limitations prevent analysis	Production centralization not universal: in some areas artisan output effectively exchanged for collective rice via a workpoint allocation. Production centralization also dependent on output control issues.

All generated a demand for coal to fire the kilns. (2) Other important activities that often competed with the state for raw materials produced within the "collectivity." These involved goods such as mats and carpets, and the making of hats. (3) Other areas. These included such traditional activities as carpentry and blacksmithing as well as the use of modern means of production such as mills to carry out contract work. They also encompassed service industries such as haircutting. Their existence was basically historical in origin and tended to reflect traditional local skills and resources. Their absorption into the cooperative often resulted from wider problems in controlling the population.

Experiences

In some cooperatives the subsidiary branches generated a relatively large proportion of output and net income, absorbing hundreds of workers. A rapid expansion often occurred in bricks and tiles, where the kiln itself was made of brick, the clay came from the earth upon which it stood, and the coal was comparatively accessible from the mines in Quang Ninh province by sea and river transport. The kilns were one of the landmarks of the delta landscape.

Many cooperatives in the sample had substantial areas under jute. This had to be processed before use, which further increased labor needs. The most important line was mat production. With hats, mats were one of two basic consumption items, the one being used to sleep on and the other the traditional conical hat worn by women.[2]

Surprisingly small proportions of output were sold to the state, and although the level of development of these activities was usually quite low they often represented the major source of cash income for the cooperative. In addition they were a focus for hopes of both absorbing superfluous labor and introducing small-scale mechanization, the latter as a visible part of the "scientific and technical revolution" (see chapter 4).

The historical origins of these areas of activity within the cooperative not only throw light upon the processes of cooperativization and the motivations of cadres but also reveal the extent of wartime disruptions to the state distribution system. In Van Tien, for example, after cooperativization the population used to "turn up for the harvest and then go away again." Nearly 80 percent of the population were Catholic, and "reactionary elements" were not always deal with "properly" by a weak cooperative leadership.[3] Before the revolution the commune seems originally to have been heavily artisanal,[4] relying upon sales of such products to finance purchases of staples from outside the commune. The indicated man-land ratio in the commune was very high. This led to difficulties after cooperativization. Collectivization of the artisanal lines was justified as follows:

As the cooperative was unable to manage and use [its] labor it was realized that the labor in the subsidiary branches had to be collectivized. Collectivization, above all under the strength of the collective, would allow the cooperative to create continuous and appropriate work for such labor. It would guarantee development of the subsidiary branches and the needs of the cooperative and the cooperators in production as well as consumption. And a necessity was to increase the income available to the cooperative. The development of production that would be allowed by collectivization and the resulting management of labor would basically result from the availability of a labor surplus for the cultivation branch. . . . This was realized in 1968. (pp. 4–5)

The ambiguities of this quote suggest the existence of considerable underlying conflicts of interest. The next-to-last sentence is revealing in the light of Van Tien's subsequent reputation as an efficient rice producer. Collectivization of the subsidiary branch activities in this cooperative followed three paths. A mass movement initially encouraged people to join the cooperative; after this the Management Committee tried offering material incentives to encourage collective participation; finally, the cooperative started up a number of new lines. These grew rapidly, and employment in the branch rose from 930 workers in 1970 to 1,550 in 1975. But with each cooperator on average working only sixty-five labor-days a year, annual incomes were still primarily dependent upon cultivation. Despite this, the reported accounting value of the labor-day was significantly higher than in rice cultivation, and members of the branch were freer to seek work outside the collective sector. This probably encouraged cooperator participation.

Van Tien used two forms of organization. Where labor could be centralized there was a III-point contract. For example, in brick-making a contract signed between the brigade and the cooperative was then subcontracted out to families who supplied a certain number of moulded bricks to the kilns in return for workpoints. Noncentralizable lines used a "putting-out" system (contracting directly with families); most of the mat-weaving was managed in this way using privately owned frames. In some of the minor lines the workers were simply buying points and the right to work: "A man working a forge has to deliver 3.0 dong each day, for which he receives 11 points, 'working' 26 work-days each month" (p. 12). This amounted to 78 dong a month, which was quite high by the standards of urban workers and cadres (typical monthly incomes, 40–80 dong). He was effectively exchanging these earnings for the cheap rice to which his points entitled him. This is a simple but clear example of the effects of the Management Committee's control over output distribution.

An essential factor in the successful development of subsidiary branches in Van Tien was the availability of raw materials. The cooperative had its own transport for coal from Ninh Binh and could obtain supplies of jute and reeds from the state. But the transport was very expensive and the state did not fulfill its contracts. Attempts to obtain more wood for the carpenters were being restricted: "The carpenters of the cooperative are usually short of work and therefore do something else[5] or go to work elsewhere; this is because there is a shortage of wood, because the team that collects wood does not obtain enough. Furthermore, the state will not allow any more people to collect wood, thus shutting off this potential source. . . . The contracts that the cooperative signs with the state regarding the supply of raw materials are never realized, because the state does not supply enough" (pp. 16–17).

As a result of the jute shortage, workers were only able to work fourteen or fifteen days a month and "had to be sent off to work in the cultivation branch." The consequences of raw materials shortages were also clear from the experiences of Giao Tan, where attempts to revive now-defunct traditional areas had failed for precisely that reason. The basic problems and conditions of these branches were relatively clear: if the raw materials were available it was comparatively straightforward to organize activities using some form of simple contracting with finance largely derived from the cooperative's control over staples. Because of the ever-present attraction of the free market it was necessary to provide a sufficiently high reward to encourage sales to the collective or participation in collective labor. This was not very difficult. But if the cooperative could not secure supplies of raw materials this problem was irrelevant:

> It is essential to solve properly the problem of "voluntary and mutually beneficial" contractual relations. The state supplies raw materials in adequate quantities, punctually and at a rational price; the cooperative delivers a volume according to the contract.
>
> Apart from this, it is necessary to solve well the problem of puchasing jute. In conditions of continuing labor surplus in the cooperative, it is uneconomic to transport raw materials to somewhere else rather than using them to open up new branches/lines and give work to the cooperators . . . this would not only increase employment in the cooperative but would also reduce transport and supervisory costs, and bring back work to the jute-preparing branch here.[6]

It is not unexpected that this cooperative should have resented entering into contracts with the state at unfavorable prices that were not subsequently honored, while it had to continue delivering raw materials it could well use itself. Nevertheless, some cooperatives seemed to retain considerable autonomy in deciding on the allocation of such raw materials as well as final products. They often managed to retain appreciable proportions of output for use within the cooperative. In Van Tien, 1.4 m out of 4 m of bricks produced were sold to the state, compared with only 25,000 out of 240,000 tiles and 25 out of 150 tonnes of jute.

In these cooperatives demand for such artisanal products was high:

> In a cooperative that has just been set up like Giao Tan, the machinery base is basically missing. Because of this, the hand-tools and ordinary means of production are very important. At present, many implements such as ploughs and har-

rows are still lacking, and especially many other necessary items are missing, such as warehouses and drying yards; pigsties are also inadequate. . . . Materials for basic construction: bricks, tiles, limestone . . . are also lacking (with regard to bricks alone the cooperative can only meet 30–40 percent of common demands each year). The situation of broken tools is extremely serious, because breakages were often not seen to in good time, and this had a great effect upon sowing and harvesting. Besides the demands of production there are the demands of the cooperators. Their houses need repairs and rebuilding, and household utensils are frequently broken or missing. The cooperative has not set up a forge, and so cooperators have to walk 3–4 km to Giao Yen to have knives fixed, etc. . . . which loses a lot of time. Each year the cooperators alone need 300,000–400,000 bricks, which is only met to 15 percent by the cooperative. In Giao Tan, where the land area is very limited and labor is abundant, the organization of subsidiary branches to find work for the cooperators has become a pressing problem.[7]

Most cooperatives organized production in a relatively decentralized way, often under the umbrella of a brigade. Serious attempts at centralization only tended to occur in model cooperatives, although Dai Thang (Quynh Luu)—where discipline was generally very tight—was moving in this direction. But despite the existence of brigades in lime, quarrying, bricks, tiles, hat-making, and so forth, all "divided up into groups by link," productivity and incomes were still low, and "because of this the subsidiary branch has good conditions for development but has not yet the resources to absorb labor and—on the contrary—workers in the branch do not yet really make efforts."[8]

Conclusions

The factors determining the ability to increase collective production levels were similar to those in other branches. The Management Committee's ability to obtain resources from the state was often important in overcoming shortages of raw materials within its own area of control. This had strong implications for the need to extract surplus for mobilization outside the cooperative. The Management Committee's control over collective output was often a crucial factor in allowing it to obtain subsidiary branch products from areas where it had no direct control over production. Autarkic development was possible given suitable initial conditions, even if this took the form of exchanging workpoints for products generated in the household sector.

The evaluation of strategies of production centralization remains extremely tentative. At existing levels of technology it appeared that such policies had their prime effect upon the increased capacity given to the Management Committee to prevent expropriation of collective product as a result of its closer control over production. But this judgment can only remain preliminary so long as there are insufficient data to allow for a calculation of the various interactions involved. The consequences of production centralization for resource allocation are complex: it affects not only the surplus controlled by the Management Committee, but also the way in which it is utilized. Funds used to encourage labor participation may have to come from collective incomes in other sectors (e.g., from the cooperative's rice-producing "general" brigades). In the absence of adequate measures of the value put by cooperators on subsidiary branch consumption goods the incentive effects of such a shift in collective production cannot be fully estimated. And in practice, production centralization inevitably tended to reduce the pressures acting upon local cadres to take into account the free-market alternatives and price ratios confronting cooperators. In a manner similar to collective pig-rearing, vertical integration was one way of overcoming the problems arising from the use of shadow prices that differed widely from those on the free market.

A crucial statistical element to a deeper understanding of the growth path accompanying different strategies would be data on labor productivity and sectoral incomes. The sources, however, do not provide sufficiently reliable and well-defined data to allow a comparison of labor productivity in different cooperatives and different branches of production. The statistical distinction was made between "gross output," "income," and the "value of the labor-day." In practice, there was some confusion with the value of gross output, which some sources used as a basis for general comparisons, and measures of income and the value of the labor-day, apparently preferred in others. The reasons for this are unknown. While income should clearly be defined as gross output less certain production expenses, it is not known what items were included in practice, nor how accounting prices were established when there was no equivalent state duty price. Substantial variation in the accounting price of manure between cooperatives has already been noted (see chapter 8), and the "value of the labor-day" may in fact be not a measure of output but primarily a distributional calculation used to establish the financial element of collective distribution. While theoretically it should be closely related to total income generation—"value-added"—in a particular area, the precise relationship has not been establishd. It is not known, for instance, how allowance is made for the use of collective capital services by rice-producing brigades (i.e., the allocation of bank loan charges).

The absence of such information is not unexpected. The sources show clearly the widespread resistance to detailed accounting procedures and the correspondingly deep penetration of the local community by outside agents which was an important characteristic of North Vietnamese agriculture. But without reasonably complete information on resource flows between branches of production it remains impossible to evaluate different growth strategies. To give one example, while it was apparently widely accepted that subsidiary branch activity was a potent source of finance because of the ease with which it generated a marketable surplus, it may well be that larger quantities of funds could have been gained by simply selling collectively produced staples onto the free market. Cooperators' preferences for subsidiary branch activity may have reflected the enhanced capacity for own-account activity that such lighter work could provide rather than higher income levels. Examples can be presented to illustrate the existence of such considerations, but evaluation of their combined effects in any particular instance requires detailed information that is unavailable.

10

Management Committee Control and the Origins of Output Gains

In a staples-deficit area food production is of great, if not overriding, importance. Rice yields in the DRV, although comparatively high by international standards, did not reach levels appropriate to the resource availability. The sources referred to many unutilized and relatively cost-free ways of increasing output. This chapter examines the interaction between resource utilization and the changing local power balances implied by the NMS.

Implications of Diversification

Discussion in the introductory chapters of this book (especially chapter 3) emphasized the wider effects of production diversification. On the one hand it was part of an economic policy aimed at higher output and resource mobilization with improved techniques. But it was also part of a social policy aimed at increasing the power of a cooperative's Management Committee. Because of this a modernizing "technocratic" management team could encounter resistance to its attempts to extend its control over production that was to some extent independent of the likely direct material benefits. Erosion of the independence of autarkic rice-producing brigades attacked an important source of local leaders' social authority. In addition, cooperators' widespread desire to maintain access to excess 5 percent land and other encroachments upon collective property could help generate support for the existing leadership. But the overall balance of material incentives was nevertheless partly dependent upon the results of managerial reforms in terms of yield increases, improved collective distribution, and the potential for sustained economic development.

This chapter focuses upon the effects of the "reformed" management systems upon collective rice production (see table 10.9). Because of the wider function of the NMS and its role in increasing the social authority of management teams throughout North Vietnamese agriculture, the sources had to seek justifications for its use in situations where the management team already possessed sufficient authority. But in some instances such additional social authority could play a positive economic role by improving the ability of technocrats to control resources and take the collective economy onto some sustainable growth path. This complicates the discussion.

Brigade Structures

The main mechanism for controlling production in the cultivation branch was the creation of a system of subunits responsible for various parts of the production process. Table 10.1 gives details of these systems for some cooperatives of the sample. The livestock branches are included because of their close links with cultivation.

Comparison of the complexity of the brigade-team structure of a cooperative with its economic performance measured in terms of gross output per capita and growth in rice yields showed no close relationship between the two. Van Tien, whose brigade structure had seen only limited development, had a high per capita gross output but comparatively slow growth in rice yields. Dai Thang (Quynh Luu) had four specialized brigades in the cultivation branch but was also experiencing comparatively slow yield increases. It should be noted that Van Tien had an extremely high man-land ratio, but poor supplies of outside inputs and draught animals.

Diversification did, however, imply fundamental changes in economic relationships: consider Dinh Cong and the two Dai Thang cooperatives. In all three a separate brigade was now responsible for the supply of seeds while in Dinh Cong this penetration of the rice monocultural brigades had gone further, with seedlings themselves now a nonprduced input from the point of view of the brigades. All three had brigades responsible for irrigation; only Dinh Cong had no ploughing brigade, but this had been discarded "after testing."

The relationship between the degree of specialization of the brigade-team structure and the cooperative's economic performance was complex. Whereas both tended to reflect the authority exercised by the Management Committee, this authority arose in part from economic sources in the form of control over production (as reflected by the sophistication of the management system). Van Tien and

Table 10.1

Brigade Structures—Managerial Organization

Type	Vu Thang Model	Dinh Cong Model	Dai Thang (Quynh Luu) Active	Van Tien Active	Hai Nam Active	Giao Tan Active	Dai Thang (Xuan Thuy) Active	My Tho Pre-/Postreform Nominal/Active
CULTIVATION								
Fixed rice-producing brigades	8	9	5	15?	13	12	16	13/13
Day-to-day water control	1B[a]	1T	1B[a]	1B	nil	1B[a](120)	1B[a]	nil/1B
Upkeep of waterworks	1B[a]	1B	1B[a]	1B	nil	1B[a]	1B[a]	nil/1B
Seeds brigade								
Seeds	1B	nil	1B	nil	nil	1B(20)	1B	nil/1B
Seedlings	nil	1B	nil	nil	nil	nil	nil	nil/nil
Non-rice-producing brigades								
Vegetables	n/i	nil	1B	n/i	n/i	n/i	1B	nil/nil
Jute	1T[b]	nil	nil	nil	c	?	6B	nil/nil
Fertilizer brigade	1B	nil	1B	d	nil	1B(30)	nil	nil/nil
Ploughing brigade	1B	nil	1B	d	nil	e	1B?	nil/nil
Mechanical brigade	n/i	nil	n/i	d	3 groups	1B(8)	1B?	nil/nil
Other brigades	n/i	nil	nil	1B[e]	nil	n/i	nil	nil/nil
No. of registered workers	992	524	492	583	n/i	n/i	1,775?	n/i / n/i

COLLECTIVE LIVESTOCK

Pig-rearing	1B	1B	1T	n/i	2T	1B(20)	1B[f]	nil/1B
Vegetable feed	n/i	1T	g	nil	n/i	n/i	1T(8)	nil/1T[h]
Vegetable processing	n/i	1T	g	nil	n/i	n/i	1 man	nil/1T[h]
Manure preparation	n/i	1T	g	nil	n/i	n/i	nil	nil/ n/i
"Caring for the pigs"	n/i	1T	g	nil	n/i	n/i	1T(5)	nil/2T
OTHER UNITS	n/i	4T	2T	nil	n/i	1B(fish)	2B	nil/2B + 3T
No. of registered workers	ca. 160	111	92	n/i	n/i	40?	68	n/i / n/i

MEMORANDUM ITEMS

Gross output per inhabitant, dong	n/i	365	285	140	175	100	120	n/i /255
Approximate annual growth in gross paddy yield per cultivated hectare, %[i]	12	13	5.5	4.5	3	7	4.5	n/i / n/i

[a] A single brigade covering both day-to-day water control and upkeep of waterworks.
[b] The team was also active in the subsidiary branch, making mats.
[c] There were informal groups organized in the rice brigades.
[d] There were specialized teams in each rice brigade.
[e] This was called a technical brigade.
[f] The brigade was not operating under a III-point contract.
[g] Here the cooperative had organized specialized groups.
[h] The team was responsible both for growing and for processing the pigs' feed.
[i] See table 10.8.
(?)—Source not clear.

Dinh Cong provide revealing examples. In Dinh Cong the brigade-team structure was close to that prescribed in the NMS, and the pattern and growth of production reflected the activities of a highly rational management team. Van Tien's leading cadres were equally committed to economic advance but used methods far from the ideals of the NMS. The brigades were highly independent and semiautarkic, and there was no real centralization of artisanal activity (see chapter 9). Yet the cooperative had a reputation for being one of the most efficient rice producers in the district, and it effectively utilized its limited resources. In Dai Thang (Xuan Thuy) all the trappings of the NMS were present yet the results were poor, while in My Tho the system was adapted so as to facilitate the change from nominal to active status.

Part of this argument hinges upon the idea that elements of the NMS were redundant under certain conditions. When the Management Committee was sufficiently secure, results could be good even if the committee allowed rice brigades and other areas a high degree of independence. Such decentralization could assist in the attainment of managerial goals. But if the committee slavishly followed directives from higher levels in implementing the NMS and tried to regulate all activity within the collectivity, then the results could be unfavorable, particularly if it attempted to constrain the household economy. This was especially true for areas of potential conflict with the material interests of the cooperators, where even a secure Management Committee was not guaranteed success. The fine penetration of production implied by the NMS was frequently unnecessary.

Control over Production in Staples Cultivation

This section examines details of production processes and the perceived constraints on development in the reformed cooperatives of the sample to illustrate their impact on output and its development.

The output in physical terms of any product is dependent upon the available resources and the social relations that define these objects as resources and constitute the way they are used. As argued in chapters 2 and 3, analysis of these social relations should include, for instance, the privatization of cooperative assets as a factor underlying the integration of production at the level of the cooperative. But these and other issues, such as the importance of kinship relations and the deep-rooted alliances behind the problems of localism and conservatism, cannot be analyzed here—it was not known who was related to whom, nor how the interaction between norm fulfillment and collective incomes occurred in practice. The full range of social relations determining output levels are not covered by the materials. But the sources still threw considerable and valuable light upon both the rationality or irrationality of management methods and problems of resource mobilization and production development.

The sources tended to divide these problems into two areas, based upon the idea of the "link" or task (khau*). This was a segment of the labor process. The concept was a crucial element of the "scientific" analysis of labor and production processes. Dominance and control ("determination") of individual tasks raised issues of labor discipline while dominance over a system of tasks raised issues of labor allocation.

Labor allocation should, according to the principles of the NMS, have occurred "at the level of the cooperative." This phrase referred to the overall structure of labor use within the collectivity—for instance, production by specialized brigades of inputs to be utilized elsewhere and mobilization of resources from within the household economy. Thus attention was focused closely upon the form of managerial organization below the Management Committee and the level at which labor allocation occurred. On the other hand, questions of improving labor discipline involved the division of processes of production into discrete tasks or groups of tasks. Because this altered the nature of the work the use of norms could then have a great effect upon labor discipline and productivity. The problem of labor discipline was in essence that of ensuring that the results of decisions concerning labor allocation were as intended.

Potential Sources of Output Increases

The sources tended not to differentiate between areas that involved changes in technology and those that did not. They identified the following as potential measures for improving economic performance:
Fixed capital
  1. Land leveling and field restructuring
  2. Irrigation works (long-term)
  3. Introduction of new tools (e.g., hand carts)
  4. Long-term work on soil fertility (e.g., liming)
Working capital
  5. Water-works (short-term; e.g., irrigation per se)
  6. Fertilizer produced in the cooperative—manure and azolla
  7. Changes in rice strains
Planned scale and composition of output
  8. Changes in output mix—use of nonrice staples
  9. Changes in harvest pattern—rice only and winter cropping
Labor inputs
  10. Changes affecting labor discipline rather than labor allocation (e.g., control over brigade leaders)
  11. Changes in labor allocation (e.g., restructuring of production brigades (reallocation of land, etc.) and/or setting up specialized brigades within cultivation
  12. Cadre education (e.g., sending them away to study or to examine experiences of other cooperatives)

Discussion of point 10 revealed details of the labor process that reconfirmed the existence of considerable problems of motivation in collective work. At the same time the sources continually referred to the scope that still remained—even in these relatively long-cooperativized communes—for further advance along the lines of 1–12. This in itself was a major pointer to the shortcomings of the system as a whole and indicated the long-term and deep-rooted nature of the agrarian problems faced by policy makers. Cooperators were frequently unwilling to work for their cooperatives despite the apparent existence of a labor "reserve" and abundant scope for its use.

The basic elements of rice production were straightforward and the possible variation in the organization of labor and production clear. There were six elements:

1. Harvesting, transport and threshing
2. Land preparation
3. Provision of seeds and seedlings; transplanting
4. Fertilizer application
5. Care of the growing plant—pesticides, weeding, etc.
6. Water—drainage and irrigation

Almost all of these elements could be performed by independent and autonomous brigades, or equally by peasant households.[1] Supplies of fertilizer and water-control services were to some extent beyond the control of the brigade leadership. But fertilizer use usually remained low, and communal provision of water had been a traditional feature of Vietnamese agriculture. This reduced the value of control over them as mechanisms to extend the authority of the Management Committee.

In advanced cooperatives a gradual penetration of this sequence paralleled the progressive dominance of the brigades, element by element:

1. Increased surveillance of harvesting and other stages where expropriation of the product was threatened.
2. Establishment of separate ploughing brigades.
3. Establishment of separate seeds or seedling brigades.
4. Supply of fertilizer from the collective pig herd and mobilization by Management Committee from households.
5. Establishment of separate technical teams.[2]

Thus, ultimately, "The rice-growing brigade would only grow seedlings, care for the growing plants, and harvest."[3]

Elements of the Production Process

Harvesting: Reaping, transport, threshing, and drying. In the conditions of double or triple cropping common in these cooperatives harvesting yielded both a final product and a ploughable field. Harvesting at an incorrect time resulted in a cost in terms of both current and future yields; there had to be a close coordination with the ploughing teams so that no time was lost.[4] Thus there were two major issues in production and distribution: first, speed and timing, and second, the prevention of appropriation by the "wrong" people.

Since collective interests were involved, the latter encouraged joint supervision. This reinforced the relative willingness of the cooperators to join in the harvest and so helped reduce problems of labor discipline. It was frequently said that "everybody joins in." In Dai Thang (Quynh Luu) this extended to local soldiers[5] and students. Again, in Van Tien there was widespread mobilization: "Each year, during the harvest-transplanting period, the number of workers in cultivation rises to 1,100 people—in other words twice as many as the average."[6] But even this did not amount to the entire labor force, which was over 2,000.[7]

Machine threshers were common in these advanced cooperatives but there was some evidence of scale problems: machines were too large for the small independent brigades—"Hai Nam had two machines but they were too big—the warehouses and yards were scattered and small";[8] transport was "too rudimentary" to allow centralized threshing. This was a cooperative with highly independent brigades.

Advances in labor productivity attained in model cooperatives were emphasized—in Dinh Cong the work was split up and closely supervised, with favorable results (see chapter 4).

Ploughing and harrowing: Problems of draught power. Together with the planting of seedlings, ploughing and harrowing were seen as "determinant."[9] It received great attention because although mistakes could have a comparatively great effect upon yields a cooperative could not use sheer weight of numbers to get good results. The resulting need for rational methods of labor motivation was accentuated by a number of factors: the shortage of livestock; the importance of proper soil preparation; and the need to coordinate with harvesters so as to avoid leaving the fields or animals idle. Speed was all-important. Cooperators had to prepare the fields in a time-slot of only a few days in the more advanced cooperatives. The time required depended upon the crop.

There was considerable interest in this area, and the desirability of ploughing brigades was a controversial subject.[10] In practice the acceptability of this option depended upon wider issues concerned, at root, largely with struggles for dominance between rice brigades and Management Committees. Some cooperatives prematurely established ploughing brigades as a result of "enthusiasm," leading to difficulties in coordinating with harvesters.

There were great differences in technical methods. In Van Tien cooperators largely did the work by hand, probably because of the shortage of draught animals and the relative abundance of controllable labor in a tightly managed cooperative. In Vu Thang, at the other extreme, 50 percent of the ploughing was done with machines.

The ploughing and harrowing capacity of a cooperative was often tested by

Table 10.2

Draught Power and Ploughing Methods, 1975

Cooperative	Ploughing brigade(?)	No. of draft animals	Sown ha. per head	Comments	Ploughing methods
Vu Thang (M)	Yes	n/a	n/a	n/a	50% of work mechanized
Dinh Cong (M)	No	n/a	2.0	n/a	Some machine-ploughing but not in rice cultivation
Dai Thang (Quynh Luu) (ASM)	Yes	168	1.9	Inadequate	25% of work done with feed machines
Van Tien (ASM)	No	93 (1973) 84 (1976)	4.6	Chronic shortage	25–33% of work done by hand
Thang Long (ASM)	Probably	n/a	n/a	n/a	Probably extensive
Hai Nam (ASM)	No	190	1.7	Adequate	No machines reported
Giao Tan (AWM)	No	150	2.5	Adequate	No machines reported
Dai Thang (Xuan Thuy) (AWM)	Yes (temp.)	253	1.9	Adequate	No machines reported

Source: See bibliography, table B1.

(?)—Source not clear.

changes in cropping patterns. Introduction of a winter crop[11] and the associated reduction in time available for land preparation (see chapter 1) could place considerable additional demands upon draught animals and machines. In some instances the sharply increased productivity offered by machines seemed to present the only viable solution to what was often seen as an essentially technical problem.

Table 10.2 shows that animals—water buffalo—remained the dominant form of draught power. In practice they were not generally used for transport and were not directly involved in the harvest. As a result their control was not directly relevant to output appropriation—unlike, for instance, control of a drying yard. So long as the brigade leadership was unconcerned with control over the production process, as would be expected in a weakly managed or nominal cooperative (see chapter 3), they were little interested in the precise way in which the animals were kept. In such situations, however, the distributional implications of decisions—such as recompense to whomever was responsible for the animal—would presumably still have been settled within the brigade. But once control over the animals became an issue in the conflict accompanying managerial reforms and attempts by the Management Committee to control cooperative resources, then questions regarding the use of animals became highly important.

In Hai Nam and Giao Tan the animals were recognizably brigade property. The former had an established price for the hire of a ploughman and his animal (chapter 7), while in the latter the water buffalo were "put out to the brigades."[12] The phenomenon of brigade property in animals was also seen in prereform My Tho. Families usually looked after the animals,[13] and they received extra workpoints for doing so (220–250 marks in the case of Hai Nam).

The table shows both that cooperatives ploughing with machines did not always put them in a separate brigade (c.f. Dinh Cong) and that those not using machines could still have a separate brigade (c.f. Dai Thang [Xuan Thuy]). Cooperatives with relatively independent brigades (those in the AWM category and Van Tien) tended not to have permanent ploughing brigades, while all the other reformed cooperatives had them—(apart from Dinh Cong, where they had been tried and rejected).

The three cooperatives without ploughing brigades were the most interesting subgroup. In Van Tien a chronic shortage of livestock meant that much of the ploughing was dependent upon human draught power. A functional division into teams within brigades was a clear example of adaptation of the NMS by a management team in a position to do so.

> In each production brigade, there are active [*sic*] teams (ploughing, fertilizer). At any time, the cooperative is able to form the specialized teams together into specialized brigades, and control them at the level of the cooperative. However, *because of the particular conditions encountered the cooperative very rarely has to do this*. These conditions are, with regard to ploughing teams: because of the

plentiful supply of labor and despite the severe shortage of draught power, the brigades ensure that the work (ploughing and harrowing) is done in time.[14]

In a cooperative with a reputation for efficient rice cultivation and a secure Management Committee a "plentiful supply of labor" could be seen as a factor aiding efficiency where elsewhere it was often said to hinder effective use of resources (chapter 7). Something like 50 percent of the work was done without draught animals;[15] brigade independence coexisted with efficient management.

In the weakly managed cooperatives norms still governed the area a cooperator was expected to cover in a day: these were 3 sao a day for ploughing in Dai Thang (Xuan Thuy) and 3.3 sao a day for ploughing, 2.3 sao a day for harrowing in Hai Nam.[16] In Dai Thang (Xuan Thuy) the Management Committee "closely monitored" the ploughing brigade.[17] It used "strong, experienced workers" to do the ploughing immediately after the harvest.[18] Such attention was unexpected in a cooperative experiencing problems of labor mobilization (see chapter 7). In Hai Nam the ploughing norm was reported to be around eight years old and based upon only a six-hour day.

At existing levels of output and technology there were no major problems with the system of brigade-based ploughing used in Hai Nam, although considerable slack in the system meant that forty-one days were needed on average for each animal to complete work that should have only taken thirty-four days. The source attributed this problem to "bad management."[19] No difficulties were expected in expanding the winter crop[20]—at even 30 percent of the total available area, ploughing needs were still far less than those in rice cultivation. But it was felt that the trade in ploughing services between brigades (see chapter 7) should be abolished—"the cooperative should interfere, and set up a specialized brigade," while in addition the norms were "too low." The source's dislike of Hai Nam's methods was also apparent from the discussion of mechanization. Four reasons were given: (1) Mechanization of ploughing would free labor for irrigation work; (2) the water buffaloes were as expensive as workers, at 220–250 marks;[21] (3) mechanization would also have a strong effect upon other branches, "encouraging new forms of organization and the breakdown of the present division of labor"; and (4) ploughing was the easiest sector to mechanize.[22]

Because the cooperative had a comparatively high labor-land ratio, without any indications of a labor supply shortage, the first reason reveals labor mobilization problems and suggests that only a limited part of the cooperative's labor force was subject to the Management Committee's control. There is no hint of the idea that higher collective incomes might increase collective participation after mechanization. The third reason reflects a possible awareness that specialization could also help the Management Committee and weaken the rice-producing brigades. Finally, the second reason suggests that effective brigade possession and control of water buffaloes allowed them to extract a high payment for their use, limiting the Management Committee's control over output distribution.

Giao Tan, where one brigade had used its control over ploughing to hide points and reduce the surplus available to the Committee (see chapter 7), reorganized ploughing during 1975–76. The reform created a cooperative-wide system of norms that effectively ended "work in brigades, eat in brigades" and set up a land-preparation brigade responsible for ploughing.[23] By 1976 this had developed to the point where the animals were put out to the production brigades for use in ploughing and harrowing and spent the rest of the time in a specialized brigade doing irrigation work. This was clearly a transitional form. Giao Tan had been severely criticized for excessive distribution to cooperators in 1975.

In these cooperatives, the sources give no real evidence that lack of separate ploughing brigades constrained production. They even state explicitly that there were no real problems in Van Tien and Hai Nam. The analytical rationale advanced for such brigades was not convincing; indeed, in the active cooperative, Giao Tan, the circumstances of their creation were consistent with the general hypothesis that the change was part of a deliberate attempt to shift power from the brigades to the Management Committee. It was therefore particularly interesting to see such brigades rejected as inefficient in the model cooperative, Dinh Cong, unless there were machines to help justify them.

Seeds, seedlings, and transplanting. The seedlings needed for transplanting came from seed-beds where they had previously been planted as sprouted seeds. There were three possible forms of organization: (1) All the work could be done by the rice-producing brigades; (2) the seeds could be supplied to the rice brigades by a specialized group or brigade; or (3) the seedlings could be supplied by a specialized group or brigade.

In practice option 3 was not used by any of the cooperatives in the sample. All cooperatives had a seeds brigade except for Van Tien and Hai Nam where the rice brigades did all the work (see table 10.5). Many cooperatives had introduced new strains.[24] The specialized brigades helped maintain quality and carry out selection and experimentation. Detailed information about this is fragmentary: Hai Nam was apparently unsuitable for the new strains, while Van Tien again appeared unusual in that it had introduced new strains without setting up a seeds brigade.

Seeds brigades were organized in a similar way to the rice brigades but often had favorable allocation of land, technically qualified labor, and other resources. In Dai Thang (Xuan Thuy) the seeds brigade had eighteen hectares of "good land" with low acidity and advantageous conditions for irrigation and fertilizing:

> New strains require technical cultivation and considerable outlays (especially fertilizer) but they reduce the growing time . . . and give a high yield. . . . Simple experience shows that if this work [ensuring that seeds are of good quality and properly used] is done well it can raise yields by 50 percent. In reality in Dai Thang the role of new seeds has been established and is being organized along the lines of large-scale socialist production.[25]

Table 10.3

Yields from Different Rice Strains
(Paddy, Dai Thang [Xuan Thuy], tonnes/ha)

	New strains	Old strains
1974	6.38	4.38
1975	4.35	4.00

Source: Nguyen Bich Huong (1976: 8).

Table 10.4

Yields from Different New Rice Strains
(Paddy, Dai Thang [Xuan Thuy])

Type 661	8 tonnes/hectare
Agriculture "8"	7.2 tonnes/hectare
Agriculture "22.23"	5.1 tonnes/hectare
Spring rice	approx. 5 tonnes/hectare

Source: Nguyen Bich Huong (1976: 8).

The data on yields illustrate the advantages and limitations of the new strains (table 10.3). Note the yield stability of old strains in the bad weather of 1975. There could be wide variation in yields (table 10.4).

But in this cooperative the seeds brigade was organizationally weak; the quality of the seeds was lower than was feasible and there was occasional confusion: "Sometimes it is necessary to take seeds from the brigades, wastefully; half of the seeds from the Vth month crop in 1976 had to be given to the livestock because they did not sprout in time."[26]

Since the new strains required better control over water and higher fertilizer inputs the area over which they were used was often limited: in Van Tien there were difficulties with low-lying, flooded, and hilly land; Giao Tan faced a shortage of fertilizer,[27] and Hai Nam's land was too acidic.

The use of faster-growing seeds to obtain more growing time for other crops was common, and basic to the strategy of the NMS. Thang Long cooperative exemplified this, especially in the use of machines and other modern inputs to overcome the associated problems of timing and plant care.[28]

In summary, seeds brigades do not seem to have generated conflicts to the same degree as other forms of specialized organization. Although there is no direct evidence to explain why this was so, one contributing factor may have been the low net total expected costs that accompanied the introduction of new seed varieties.

Table 10.5

Use of New Strains and Methods of Organization

Vu Thang (M)	Seeds brigade set up in 1968. Area sown with new strains rose from 3% in 1965 to 61% in 1969 to 100% in 1970.
Dinh Cong (M)	Separate brigade responsible for seed and sprouting. Groups within each production brigade used for pulling seedlings, transporting, and planting out.
Dai Thang (Quynh Luu) (ASM)	Seeds brigade had 32 workers and was similar to rice brigades but "about half the size."
Van Tien (ASM)	No seeds brigade. All of the better land (80% of total area) had new strains; remainder was hillocky or low-lying and used old strains.
Thang Long (ASM)	New seeds were "fully introduced" over the period 1967–1974.
Hai Nam (AWM)	Cooperative could not introduce new seeds because of high soil acidity. No seeds brigade.
Giao Tan (AWM)	Seeds brigade had 20 workers and 7.6 hectares; in 1974, 67% of land had new strains; in 1975 this rose to 70%.
Dai Thang (Xuan Thuy) (AWM)	Large seeds brigade with 30 workers and its own land (8 hectares) responsible for providing seeds to all 16 production brigades (ca. 110 tonnes annually). Area sown with new strains reached 45% of total in 1974, 55% in 1975, and was planned to rise to 80% in 1976.

Source: See bibliography, table B1.

Although such brigades were given technically qualified workers and presumably provided an example of the possibilities for efficient production, it is notable that they were never used as objects of emulation. This may have been because of the lack of comparability resulting from the limited area involved and the favorable resource endowments. The sources did not try to justify cooperatives that did not allow the brigades to grow their own seeds: even in cooperatives such as Giao Tan where the Management Committee had been unable to reallocate land between rice brigades there was little difficulty in obtaining land for the seeds brigade (see table 10.5). Although the two cooperatives that lacked seeds brigades were also the ones with "scattered land"—Van Tien and Hai Nam—these two cooperatives could not be used to justify their introduction. There had been no problems introducing new seeds in Van Tien while in Hai Nam the soil was unsuitable.

The widespread establishment of seeds brigades suggests that the opposition to attempts to reduce the autonomy of rice brigades was rational, in that it would be much lower whenever the expected gains were adequate.

Fertilizer: Supplies and availability. There were three reported sources of fertilizer: manure (from both household and collective pigs, other animals, primarily draught animals, and night-soil); green fertilizer (mainly azolla, a form of duckweed); and chemical fertilizer (nitrates, phosphates, lime, and potash). Cooperatives showed substantial variation in fertilizer use. The widespread failure of attempts to develop the collective pig herds led to pressure on the household sector to supply fertilizer from the private pig herd. There was no information at all on determinants of the supply of chemical fertilizer such as the state's allocation system and the possible influences on it of a cooperative's level of grain deliveries.

The predominant type of green fertilizer—azolla—was grown on the surface of water, usually in paddy fields when they had no rice in them. In the nineteenth century the techniques required had been the closely guarded secret of a few communes. Labor inputs were low, and the main constraints on production appear to have been the area of water available and the competing uses to which such land could be put. In areas that suffered from flooding the opportunity cost in terms of alternative crops was probably very low.

There was information on azolla use for Van Tien and the three weakly managed cooperatives of the sample (see table 10.6). From 15 to 100 percent of the available area was used. Since azolla was technically easy to produce there was little justification for using any form of centralized labor organization. In Van Tien, "with regard to groups producing fertilizer: pig-rearing and azolla . . . have developed very well, and so the brigades are fully able to supply themselves with fertilizer. Because of this, the fertilizer brigade rarely does anything, despite the fact that its constituents—the specialized teams in the brigades—are very active."[29]

These teams supervised the work of families on part of the land after the spring/winter harvest(s). The work was seen as an incursion into the household economy similar to the control exercised by the cooperative over the households' pig-rearing activities.[30]

The norm for growing azolla in Dai Thang (Xuan Thuy)—1 mark for 300 kg— was indistinguishable from the exchange of the product for workpoints.[31] In Giao Tan there was a "green-fertilizer" brigade in addition to an azolla brigade, dating from 1974;[32] it was not clear why two brigades had been set up.

The land in these communes had been under almost continual cultivation for centuries, if not millennia, and fertilizer was of great importance. The pattern of fertilizer allocation reveals the different relative priorities attached to crops; in practice, staples production was not always the most important. Other sources of increased fertilizer demand included the use of new rice strains; an expansion of a winter crop, especially vegetables; an expansion of "industrial" crops such as jute; and an expansion of the green foodstuffs needs of the collective pig herd.

Table 10.6 suggests that there was no simple pattern of fertilizer allocation. Although the winter crops that in some places were expanding rapidly had a strong claim on supplies, they might still be accorded very low priority. Thang Long cooperative was producing enormous quantities of subsidiary crops that were sold to the state. This was facilitated by large inputs of machinery and chemical fertilizer, supplemented by manure from the collective pig herd.[33] In Giao Tan the cooperative had set aside half of the available manure in the winter

Table 10.6

Fertilizer Availability and Use, 1975

	Vu Thang (M)	Van Tien (ASM)	Thang Long (ASM)	Hai Nam (AWM)	Giao Tan (AWM)	Dai Thang (Xuan Thuy) (AWM)
Pig manure						
Collective	n/a	250–300 t	n/a	40 t	n/a	112 cu. m[a]
Household	4,000 t	4–5,000 t	n/a	3,400 t	n/a	6,140 cu. m
Total	9,760 t[b]	5,000 t	n/a	n/a	n/a	6,300 t
Other sources of natural fertilizer						
Night-soil	n/a	n/a	n/a	n/a	n/a	Used, quantity unknown
Azolla	n/a	2,000 t; 15% of harvested area	n/a	30–50% of cultivated area	100% of riceland	Used, quantity unknown
Chemical fertilizer	n/a	n/a	400–500 kg/ha[c]	165 kg/ha[d]	60–80 kg/ha[d]	Ca. 500 kg/ha[e]
Reported organic fertilizer use on collective riceland	n/a	11–12 t/ha[f]	8–10 t/ha[f]	n/a	5–6 t/ha[g]	10 t/ha[g]
Apparent organic fertilizer availability	34 t/ha	33 t/ha	n/a	7.5 t/ha	n/a	12.5 t/ha

Source: See bibliography, table B1.

Note: Unless otherwise stated, all data are in crude "unadjusted" units per cultivated hectare per year.

[a]*Sic*—the source gives quantities in cubic meters.
[b]All organic fertilizer.
[c]Nitrogenous fertilizer per cultivated hectare.
[d]The level of supplies from the state was quoted as being 244.5 tonnes, of which 141 tonnes were "nitrogenous (*dam*). This was described as "a level not enjoyed by many other cooperatives."
[e]This was made up of 160 kg/sown hectare of nitrogenous fertilizer (*dam*), 150 kg/sown hectare of "phosphates" (*lan*) (source not clear), and over 100 kg/sown hectare of lime (*voi*).
[f]Reported application per sown hectare.
[g]Reported application of manure per hectare of riceland.

crop of 1976 for growing potatoes in order to expand feed supplies to the collective pig herd.

In Hai Nam jute yields were falling because the small amounts of chemical fertilizer allocated to the cooperative were inadequate to maintain soil fertility; there was no winter crop. Calculation showed a potential of around 18 tonnes per hectare of fertilizer (including green fertilizer,[34] collection of the draught animals' manure, etc.) compared with an actual usage of 10 tonnes per hectare. But only 30–50 percent of the possible area was used for azolla despite the fact that in earlier years the full potential of 80 percent of the paddy area had been attained. No reason was given for this astonishing decline.[35] The familiar problems encountered elsewhere in mobilizing cooperative labor were present in both Giao Tan and Hai Nam.[36]

When discussing the situation in Dai Thang (Xuan Thuy), the source clearly ranked fertilizer supplies below water control: "In a situation where the cooperative has basically solved the problem of water control it is fertilizer that becomes the most important factor."[37] Given the existing production targets in this cooperative the implicit fertilizer "shortage" was nearly 5,000 tonnes in 1976: "To resolve this difficulty, the cooperative can utilize its level fields immediately and grow azolla on 200 mau [72 hectares] of winter land. If we take the norms of 300 kg/mark, and one worker each year working 240 marks [days], then this corresponds to 100 workers [10 percent of the reported labor surplus]."[38]

The sources do not permit firm conclusions to be drawn on fertilizer use. Table 10.6 shows that both weakly and strongly managed active cooperatives appeared capable of generating high levels of fertilizer supplies. The detailed evidence from the texts suggests that considerable scope for increasing fertilizer availability was not always realized. Both animal manure and duckweed were possible sources of additional supplies.

Deliveries of chemical fertilizer from the state were in some instances appreciable (e.g., the weakly managed Dai Thang [Xuan Thuy] and the mechanized and vegetable-exporting Thang Long). The sources did not reveal the factors determining supply levels.

Fertilizer use frequently appears to have concentrated upon nonrice crops to a greater degree than might be expected from the overall attitude of the NMS toward the interrelationship between pig-rearing and rice cultivation. This was sometimes associated with development of the collective pig herd (see chapter 9). Labor problems encountered in the collection and preparation of fertilizer were similar to those encountered elsewhere, and they were more prevalent in weakly managed cooperatives.

Care of plants. In the conditions of high humidity and continuous cultivation faced by farmers in the North Vietnamese deltas, and especially with the introduction of new strains, pests and disease were particularly acute problems. Labor requirements for pesticide application were not great, and its organization did not pose problems. Teams of some sort were in use in Dai Thang (Xuan Thuy) and

Giao Tan, whereas in Dinh Cong each fixed production brigade had its own group.[39] The only other mention of pesticide use was in Thang Long, which had five sprayers. The labor demands of the other aspect of plant care—weeding— were so minimal that (perhaps mistakenly) almost no mention was made of it; the indications were that the work was done in slack periods by members of the fixed brigades.

Day-to-day water control. The work needed to ensure that the irrigation and drainage system was providing a correct water level in the fields was separate from development and maintenance work. Because of the extremely unstable level of precipitation, cooperatives frequently organized large-scale labor mobilizations to cope with flooding and drought. Flooding was usually the more serious problem.

Day-to-day work seemed usually to be carried out within the rice-producing brigades, often supplemented by some form of organization at the level of the cooperative. Sometimes this took the form of a separate brigade that also took responsibility for maintenance and development of the system. Such brigades had a variety of names.[40] In Van Tien there were two brigades of this type, one of which was controlled by the district.[41] This helped integrate the commune's system into that of the district.

The rapid mobilization of labor to "save the crop from flood" not only required the participation of people who would not usually have worked for the collective (i.e., the full labor force of the cooperative rather than those "permanently" employed) but also involved a certain degree of altruism if the brigades whose crop was not threatened were to participate willingly, unless the cooperative had established "unified distribution." It was therefore hailed as a classic example of idealized cooperative labor indicative of a "strong socialist spirit." Such an incident occurred during the Xth month crop (1975) in Giao Tan. Brigade-level distribution was not abolished until the Vth month harvest of 1976, but part of the cooperative's land had been reallocated to brigades in 1971.[42] A low-lying field was flooded and there were only three days in which to drain it. The pumps[43] and the labor force were inadequate: "only when all available labor was used could the harvest be saved."[44] The exercise increased output by 54.6 tonnes, but no mention was made of any special remuneration: "This is the real proof of the necessity for enlarging the scale of cooperativization in Giao Tan."[45]

The prices used in cooperative accounting systems implied that the imputed water costs were relatively low. There is no information as to the possible inclusion of capital or investment costs in these material expenses of production. In Van Tien the itemized costs of paddy for 1975 included 3.6 dong/tonne for irrigation and drainage plus 2.4 dong/tonne for antiflood work among a total of nearly 140 dong for material expenses.[46] The sources did not record any allocation of such accounting measures to the suppliers of such services.

Casual observation of delta farmwork showed that much primitive and unmechanized water work was relatively small-scale. Children carried out a substantial proportion of it.

Only the direct control of water has been examined here. Direct investment in dykes and water channels is considered in the next section.

Land Improvement and Hydraulic Works

Direct investment by a cooperative of a collectivity's labor resources in infrastructural and other projects was one of the main justifications advanced for agricultural collectivization. The sources placed great emphasis upon collective activity to level land, improve drainage systems, and so on. The extent to which it was possible to find projects of this type tends to confirm the limitations of the old communal system.[47] In assessing the value of these projects, however, there were two complicating factors: first, collective labor was conventionally deemed to be good per se; second, a frequent justification given for leveling land, increasing field size, and so on was the need to prepare for machines which in the event did not arrive.

Despite the resulting tendency to overestimate the net value of such projects, they could result in large yield increases if the water control system allowed the introduction of additional rice crops. Better irrigation and drainage were also vital to the yield stabilization required to reduce uncertainty and therefore facilitate planning. Introduction of a nonrice winter crop not only needed better water-control but also necessitated the use of riskier but faster-growing rice strains.

The social viability of these investment projects is susceptible to approaches familiar from standard economic theory. It is analytically important to distinguish between two types of problem. First, difficulties arose in convincing cooperators that they should participate in the short run when they would only benefit in the long run (or so they were told). Second, there are the further problems created by the distributional and wider social implications of a project. Not only will different groups often have divergent interests, but power relations could be affected by wider consequences (e.g., the level of taxation, privatization of collective assets, brigade leaders' ability to siphon off collective output onto the free market). Because of such considerations, cooperators and brigade cadres might see it as in their interests to oppose cooperative investment projects. This could occur despite the apparently beneficial effects of such activities upon total net output. In such situations a "game" with a potentially positive social sum could be perceived (possibly correctly) as having a local negative sum. People who have lived within such a social system for over two decades are likely to have acquired strong opinions concerning the predictability of outcomes and the value of existing methods of insuring "contract fulfillment."

In practice the commune appears to have remained an important level for the perception of group interests. Amalgamation of cooperatives to the commune level therefore corresponded in part to a possible congruence of production and distributional units with underlying social groups capable of resolving the problems of collective decision-making discussed above. But, as has been repeatedly emphasized, cooperatives existed within a wider collectivity and were themselves divided by potential conflicts between Management Committee, brigades, and cooperators. The mobilization of labor into projects that appeared—especially to the authors of the sources used here—to confer considerable long-term benefits therefore often encountered "irrational" difficulties. Policy implementability was not guaranteed, and the feasibility of the various suggestions put forward greatly depended upon local conditions.

Table 10.7 collects information on the work that had gone into land improvement and waterworks. Most of this difficult labor was done by hand, and mobilization of people to do it was a profound test of the powers and abilities of local administration, for it was well known that machines were available for the fortunate. But machines were also expensive: "Experiences with land improvement in many places tell us: if done by machines it is very expensive—around 150,000 dong per hectare, while the capital of the cooperative [Dai Thang, Xuan Thuy] is still very small, and labor very abundant. Because of this one has to mobilize and encourage labor not yet fully utilized in the cooperative to participate in improving the land. This changes living labor into materialized labor, creating the material basis for accumulation in the cooperative."[48]

Few cooperators seemed enthusiastic about the work involved. The cooperative was still comparatively poor (a per capita distribution of ca. 15 kg per month), and the planned projects required 360,000 man-days. The work was to improve 110 hectares of water-logged rice land and therefore raise the yield from 3 to nearer 6 tonnes per hectare. But in its three years of existence the cooperative had only finished 11 hectares, although it was "well aware" of the desirability of finishing the job. The cooperative had had fewer problems in mobilizing labor to work on the commune's irrigation system,[49] perhaps because returns to the private plots were present. During the bad weather of 1975 the cooperative did better than others in the region, and this was attributed to the new irrigation system.

The potentially dynamic role of active cooperatives in this area came out clearly in the case of Giao Tan. The cooperative was set up in 1974 from two smaller cooperatives: the commune of Giao Tan had three villages, two of which formed a new cooperative in 1969, while the third joined with a village from another commune. Before 1971 there was no real water-control system, and the existing drains between the two cooperatives were not well maintained because "responsibilities were not clearly defined."[50] Amalgamation and greater control allowed the cooperative to mobilize labor and deal with the situation. A complete new system was constructed. But the cooperative still needed to construct ten more drainage channels: "without amalgamation it would be hard to construct these, because (1) they are designed to boost production outside the area where they are to be built; (2) the drainage system could not be well-managed."[51]

Here the expanded cooperative could help overcome the sentiments of local-

Table 10.7

Land Improvement and Water Works

Vu Thang (M)	Well-endowed with machines; reconstruction of water works very important in reducing acidity and salt content of soil.
Dai Thang (Quynh Luu) (ASM)	Virgin land of 200 hectares of which 66 hectares had been brought into cultivation. Cooperative was cooperating with district on large-scale projects.
Van Tien (ASM)	Constructed complete water control system "between levels 3 and 1" over entire cultivated area. By 1973, value was put at ca. 42,500 dong. Land was ready for machines. Water-wheels in use.
Thang Long (ASM)	Had set up a large system of canals that used district pumping station and radically changed production conditions by allowing two crops.
Hai Nam (AWM)	90 hectares of land that could have been cultivated remained unused. Although there was no long-range planning by cooperative, water-control system was thorough, covering over 85% of cultivable area: 65%—pumps, 25% gravity, 10% wheels. Pumps were concentrated on rice rather than jute.
Giao Tan (AWM)	Before 1971, had no system of water control. Since then all drains redug and complete "level 3" system constructed. With combination of pumps and manual labor the whole cultivated area was covered (15% by gravity). Because cooperative owed money it had to use manual labor to dig drains, and because they would only "benefit other areas," ten priority drains had not been dug.
Dai Thang (Xuan Thuy) (AWM)	Had improved 30 of 300 mau of water-logged land. Shortages of finance apparently prevented machine use. Sixty percent of cultivable area now "well-irrigated." System had been rebuilt and pumps were now used. One constituent cooperative had not been using 20% of its land before amalgamation and was now doing so. Cooperative had "high" land that was still not used. The work done had had a good effect on the reduction of instability in yields.

Source: See bibliography, table B1.

ism that accompanied the presence of a foreign village in the same cooperative as three villages from the same commune. In Thang Long, however, the sub-commune level cooperatives had been able to cooperate on this sort of venture. The district had assisted this process by integrating them into its water-control system. The two cooperatives that eventually merged to form Thang Long had been set up in 1966 with 220 hectares and 90 hectares: "At this scale, the two cooperatives constructed a large system of canals, joining the fields to the pumping station of the district and changing their conditions of production."[52]

Table 10.7 summarizes experiences in the sample. It shows how the dynamic potential of the progressive managements typical of these cooperatives in practice ran into the enormous difficulties—of labor mobilization, shortages of means of production, and so on—that characterized North Vietnamese agriculture.

Origins of Yield and Output Gains

Table 10.8 summarizes the limited information available to assist comparison of the economic performance of the advanced cooperatives of the sample. The data on growth in gross paddy yields are derived from table 5.5. These data suggested that these cooperatives were generally attaining higher yields than the DRV average: the average for the entire sample for the two years 1974–75 was over 60 percent higher than the reported DRV average, ranging from nearly twice as high in Dinh Cong to plus 15 percent in Dai Thang (Xuan Thuy).

The two model cooperatives both showed yields well above those of the active cooperatives of the sample. Comparison of the rates of growth in rice yields tended to confirm the superiority in this area of the model cooperatives. The subgroup of strongly managed cooperatives also appeared slightly more successful on average than the weakly managed group, with an estimated growth rate of just over 5 percent p.a. in rice yields compared with around 4.75 percent for the other group. It should be noted that reported average yields per hectare of riceland in the DRV rose only 1.7 percent p.a. between 1965 and 1975 (see table 5.5). Given the likely inaccuracies in the official data such figures are highly unreliable. It can safely be concluded, however, that both yields and yield growth rates were above average in the advanced cooperatives of the sample.

Within the two subgroups of active cooperatives (weakly managed and strongly managed) some highly tentative comparisons can be made between the different growth paths. The comparative success of Thang Long in attaining a high rate of growth of paddy yields was accompanied by extensive provision of modern inputs—machines and chemical fertilizer—accompanied by a complete adoption of new seed strains and large investment in land-leveling and water-control projects. A simple explanation for the comparatively slow rate of growth of rice yields in Van Tien might be seen in the lack of provision of modern inputs, with almost no machines and no mention of chemical fertilizer. The strategy appears to have been to concentrate initial efforts upon reaching self-sufficiency in staples, after which resources were channeled into the subsidiary branches.

In the subgroup of weakly managed cooperatives the relative lack of success of Dai Thang (Xuan Thuy) suggests that without effective measures to ensure proper resource utilization, comparatively high fertilizer inputs and not insignificant investment in land improvement and water works could have only poor results. Hai Nam had no ploughing machines, low chemical fertilizer allocations, poor investment in land improvement and water works, and was also unable to introduce new seeds. It probably represented a relatively close approximation to production conditions in nominal cooperatives.

Vietnamese Explanations of Success at the Grass Roots

The question of how best to explain the different levels of economic performance

Table 10.8

Economic Performance of Model and Active Cooperatives: Staples Branch

Cooperative type	Gross output/inhabitant	Estimated growth in gross paddy yields[a]	Ploughing methods[f]	Use of new seeds[g]	Fertilizer[h]		Investment in land and water-control[i]	Population density[j]	Additional comments
					Organic	Chemical			
Model									
Vu Thang	n/a	12% pa[b]	50% of work mechanized	100%	High	n/a	Extensive	15.2	High paddy production financing collective and household pig-rearing.
Dinh Cong	365 d	13% pa[c]	No machines in rice cultivation	Probably 100%	n/a	n/a	n/a	15.5	Model cooperative without machine ploughing; outside inputs not mentioned but probably favorable.
Active (strongly managed)									
Dai Thang (Quynh Luu)	285 d	5.5% pa[d]	25% of work mechanized	Probably 100%	n/a	n/a	Extensive	9.7	Highly disciplined management system with very close labor supervision and high penalties.
Van Tien	140 d	4.5% pa[b]	1/4–1/3 of work done by hand	80% of cultivated rice-land	High	n/a	Extensive	28.4	Extensive development of subsidiary branch activity; secure Management Committee allowing some brigade independence.

Thang Long	197 d	9% pa[e]	Probably extensively mechanized	100%	High	High	Extensive	n/a	Highly mechanized co-operative producing a large "merchandised surplus" of vegetables.
Active (weakly managed) Hai Nam	175 d	3% pa[d]	No machines reported	nil	Medium	Low	Poor	17.5	A weakly managed cooperative with largely unreformed independent brigades.
Giao Tan	100 d	7% pa[d]	No machines reported	70% (1975)	Low	Low	Medium	24.4	A weakly managed cooperative; Management Committee's attempts at reform had no outside assistance.
Dai Thang (Xuan Thuy)	120 d	4.5% pa[d]	No machines reported	55% (1975)	High	High	Medium	18.2	Weakly managed co-operative; reform assisted by outside inputs, but with little success.

[a] Yields per cultivated hectare; see table 5.5.
[b] 1965–1975.
[c] 1970–1975.
[d] 1965–1975; 1965 data assumed equal to DRV average.
[e] 1965–1975—gross staples production.
[f] Derived from table 10.2; see also table 5.4.
[g] Derived from table 10.5.
[h] Derived from table 10.6; estimates are highly uncertain.
[i] Numbers per hectare of rice-land—derived from table 5.1.
[j] Derived from table 5.1.

experienced in these cooperatives also confronted the Vietnamese authors of the sources, who adopted two approaches to the analysis of yield increases. These were distinguished by the relative emphasis put upon administration and organization as opposed to changes in the physical conditions of production. This difference of interpretation reflected fundamental problems faced by policy makers and cadres in North Vietnam. Those who emphasized organization and management typically confronted situations where the occupation of positions of power and authority by "progressive elements" was in question. Once such elements were firmly established the focus shifted to consideration of how they should best use their power: thus the "technocratic" emphasis upon improved methods of production.

When the brigades were quasi-independent—as in a nominal cooperative—the emphasis was upon distributional relations and the control of output and means of production; because technology was almost static and per capita output at best unchanging, control over distribution was more important than control over production. But since cadres who wished to implement managerial reforms often lacked the power to do so, some sources stressed the need to create an organizational framework within which they could work. This was the basic origin of the organization and management viewpoint, which is consistent with the opinion that relations of production have some priority over productive forces. On the other hand, since cadres in developed cooperatives often went too far in emulating certain production forms, model cooperatives were sometimes presented as solving managerial problems by basing production forms upon technical conditions. This technocratic viewpoint in effect maintains that productive forces take some some priority over relations of production. It is particularly interesting to observe that the theoretical and practical tension between the two categories can here be related operationally to different sets of practical problems confronting policy makers and cadres.

The technocratic approach tended to place far less weight upon problems of leadership and to look instead at changes in the technical conditions of production—changes in water-works, seeds, fertilizer, and so forth. Here this viewpoint provides some additional but limited information to help examine the perceived origins of yield changes in apparent isolation from the changes in social relations that made them possible. Because of data limitations such an analysis perhaps throws more light upon the conditions confronting cadres than upon the parameters underlying the observed changes in economic variables.

The first administrative viewpoint was common among authors who discussed the initial problems faced in the creation and early history of the model cooperatives, and the contemporary relevance of the solutions adopted:

> The experience of Vu Thang demonstrates that any cooperative that wants to succeed must, before anything else, ensure that it is strongly united and must define the direction of development and methods of advance clearly and correct-

ly. For instance, it is ridiculous to maintain that when it had just been set up Vu Thang was not well aware that the basic reasons for the lack of development of production were that the fields were continually subject to floods and drought, and were uneven and acidic. In fact, the problem was that the Party Committee was not yet united in a determination to set up hydraulic works, and there was a prevalent sentiment of "How can we do it alone?" As a result nothing was done.[53]

In a nominal cooperative, "unification" of the Party Committee meant a subordination of the local party cells. These based themselves upon hamlets and villages and corresponded either to the smaller cooperatives of the commune or to the brigades of the cooperative; in extremis, it meant a subordination of the "fiefs" of the dominant members of the party to the Party Committee's leadership. But in practice the limited opportunities and likely costs of adopting the party center's prescriptions led to claims for outside assistance, as in Vu Thang (see chapter 4).

A good example of the second, technocratic viewpoint, here critical of regressive behavior (in relatively advanced cooperatives), came from the study of Dinh Cong cooperative:

In our country, advancing from small-scale production to large-scale socialist production, collective forms of labor have arisen before machines could replace rudimentary tools. Tools are still basically simple. Because of this, in order to organize and develop the brigade forms we must establish the peculiarities of the implements used and then choose appropriate forms. At the same time we must base ourselves upon . . . technical progress. If this is done we can avoid two mistakes: (1) "show-off" forms, leading to an unwieldy and imitative organization of labor that pays no attention to economic results; and (2) abolishing or weakening the fixed production brigades. These mistakes have been thoroughly condemned. . . . The material-technical base . . . and the technical level . . . in each branch of production (construction of an irrigation system, small-scale mechanization, development of buffalo carts, the technical level of new rice-seeds, etc.) *determine the development (improvement, replacement, establishment) of the forms of organization and labor.*[54]

The article went on to criticize cadres who set up independent ploughing brigades "irrationally" (i.e., for noneconomic reasons) when there were no tractors. But sufficient "economic" factors to explain such behavior are easy to suggest: the desire to obtain access to economic resources from the state (e.g., favored treatment for being a cooperative close to the prescribed "ideal," bonuses for allowing immediately superior levels to present favorable reports to higher levels) or a desire to attack "conservative" elements entrenched in the brigades by eroding brigade autarky.

Table 10.9

Management Committee Control and Origins of Output Gains: Summary of Main Findings

Issue	Time-series	Cross-section
Brigade-team structure—complexity and economic performance	Rapid growth in yields not guaranteed by establishing formal structure of the NMS	Contra-examples illustrate possible effectiveness of simple structures in efficient resources mobilization and poor results of complex structures in some cooperatives
Autarky and production independence of brigades and its relationship with the strength of cooperative managements	Rapid dominance of rice brigades in some cooperatives but erosion of production independence by use of specialized brigades not universal	Weakly managed cooperatives could have specialized brigades, strongly managed cooperatives need not

Managerial control over various stages of staples production

General	Formal implementation of the NMS intended to dominate production activity in each stage of production	—
Harvesting et al.	—	Labor mobilization in tasks associated with harvesting comparatively easy in most cooperatives

Ploughing and harrowing	Increasing attention paid to ploughing and harrowing—ideological pressures for separate brigades frequently hard to justify	Independence of rice brigades in weakly managed cooperatives reinforced by brigade control of draft animals. Ploughing brigades not universal in strongly managed cooperatives
Seeds	Few problems encountered in establishing seeds brigades	No apparent correlation between strength of cooperative management and ease of adoption of new seed varieties
Manure	Manure from collective pig herds and other sources of potential increases in collective supplies not always utilized in collective rice growing (data very poor)	Substantial variation in availability of both organic and chemical fertilizer, but with poor data no clear pattern of allocation discernible
Plant care	Caring for plants not (perhaps mistakenly) an area of concern	—
Water control	—	Extraordinary labor mobilization for flood control possible in weakly managed cooperatives
Direct investment in land improvement and water works	Many apparently profitable and viable land improvement and water-control projects still not implemented in many cooperatives but much work already carried out	Problems of labor mobilization could be overcome by the use of the wider "social authority" of a strong management team
Origins of increases in land yields land yields	Changes in social systems within the cooperatives of the sample were associated with acceleration in yields compared with reported national averages. Data inadequate to explain precise origins of changes. Subjective impression of "technocratic" Vietnamese authors that "land improvement" was most important, followed by new seeds and fertilizer	Tentative conclusion that effective resource utilization (both local and supplied inputs) dependent upon strength of management team

The technocratic authors argued that once the Management Committee could mobilize resources and make "economically rational" choices, the most crucial of the various inputs was what (literally) they called "land reform" (*cai tao dat*). This term referred to the complex of tasks (irrigation, flood prevention, field-leveling, and soil care) required for the introduction of new seeds: "If irrigation has provided the material-technical base . . . then seeds are a source of land [*sic*], inexpensive, and rapidly raize the yield of production."[55] Similar points were also made by Vu Duc Thang (1976), Tran Quang Lam (1976), and Nguyen Hoang (1979).

The areas identified above where changes in inputs were possible might now be given an approximate ranking as follows:

1. Water works and land improvement (nos. 1, 2, and 4) were the essential prerequisites.

2. New seeds were then capable of great increases in yields (no. 7).

3. Fertilizer. Despite the emphasis put upon it in official reports, fertilizer (no. 6) was typically ranked below water works and land improvement.

The sources used provided insufficient information on the relative importance of other factors to permit their inclusion in this ranking. The superficiality and inherent lack of clarity of such debates stems at root from the absence of clearly defined and isolated goals that accompanied the orthodox conception of the Vietnamese Revolution. The discussion will return to these issues in the next chapter.

Conclusion

This chapter has examined a number of issues arising out of attempts to improve the economic performance of the staples branch of collective agriculture. In the absence of detailed statistics on resource utilization it has been impossible to carry out a full analysis of the factors determining the comparative successes and failures of the different managerial systems. The chapter has attempted, however, to examine systematically the various stages of production in staples cultivation in order to evaluate the possible justifications for the erosion of the production independence of autarkic rice brigades implied by the NMS. Because of the lack of comprehensive data conclusions are highly tentative; the basic findings are summarized in table 10.9.

Part IV

The Agrarian Question in North Vietnam

11

The Agrarian Question and Cooperative Nominalization

The next two chapters look at the Agrarian Question in North Vietnam from a wider perspective. In this chapter a historical comparison illuminates cooperators' basic interests and permits some understanding of how they differed from those of top leaders responsible for policy formation. In particular, the state of direct material interests associated with the question of cooperator incentives had important implications for the possible evolution of the formal system. Here the position of local cadres and their role in the operation of the complex management structures of the NMS, especially the brigades, was a major issue that came to the fore as policy began to change in late 1979. The detailed discussion in chapters 7 to 10, revealing cooperator attitudes toward the relative value of various parts of collective activity, provides useful indications as to those areas where collective structures had a real economic basis that could gain popular support.

The effective collapse of the NMS as a viable policy in late 1979 was not immediately followed by a "reformed" set of new policies. And once the so-called output contract (*khoan san pham*) system was in place in late 1980 it was still far from universally accepted throughout the North. At the same time the authorities introduced other measures to do with procurement policy and income distribution within cooperatives, which helped form an overall package that had a substantial positive effect upon producer incentives and therefore upon output levels. But these changes essentially sought to prevent a complete decollectivization of Northern agriculture, and have to be seen in terms of a movement toward a new compromise between the aspirations of the national leaders and the rural periphery. Chapter 12 discusses these changes in detail.

Ideas and Interests

Although by 1979 North Vietnamese peasants had had nearly twenty years of experience with the collectivized system, this compares with the previous centuries of life in the corporate dynastic communes. Also, whereas the new government's commitment to independent national economic development sharply distinguished it from both the French and dynastic rulers, the basic problems of high risk and low incomes confronted cooperators as they had their forebears. The historical perspective is therefore important to understanding the nature of the Agrarian Question in North Vietnam, and the underlying compromises that it entailed.

In a risky environment such as the Red River Delta the minimum income support provided by corporate organizations will usually have high social value. And both during and after a war generally felt to be just, such organizations' abilities to pay pensions and invalid benefits will also be appreciated. These are simple arguments for the existence of some form of collective or corporate organization. They are not, however, arguments for the existence of collective production units such as producer cooperatives. To justify the involvement of local organizations in the control and direction of economic activity requires quite different considerations.

A conventional developmental analysis of a policy of agricultural cooperativization would typically raise the following issues:

—the general pattern of development of agricultural production and its compatibility with national economic development goals.

—the creation of sufficient increases in the mobilized surplus of a range of products appropriate to both the short- and long-term needs of modernization and industrialization, taking into account the demands put upon the rest of the economy.

—the change in real living standards of cooperators and others.

—the change in income distribution, both absolutely (e.g., the maintenance of subsistence floors) and relatively.

—the impact on noneconomic areas, that is, the "democratic" nature of management methods may be considered worthwhile in themselves.

—the extent to which alternative policies may have been more successful in attaining desired goals.

A key element of cooperativization policy is the assumption that sufficient economies of scale exist in the organization of collective labor in the brigades. The general conclusion of this study is that these were absent. This conclusion is weakened, however, by the effects of the overall aggravated shortage economy on the incentive pattern, which, by altering distributional relations, biased incentives against the collective. Cooperators' work on their own account effectively gained higher rewards because output there was disposed of on the free market or suffered lower deductions for taxes, local administrative costs, and so forth. If cooperatives had had the same rights of disposal as cooperators, facing similar distributional opportunities, then comparisons of the underlying scale economies would have been far easier. Expectations that introduction of a new management system would likely lead to greater procurement and deductions for local cadres would of themselves greatly offset any underlying economic advantages possessed by the new system. At the end of the day, however, it is clear that any economies of scale offered by collective production were insufficient to offset the unfavorable distributional context. The underlying sense one gets is that such gains were not usually significant. The evident ability of active Management Committees to modify the NMS away from full centralization supports this argument, as does the way in which policy eventually moved after 1979.[1]

This study has argued that the producer cooperative model was somewhat isolated from the real requirements of the local economy, which was why it was so severely modified in practice whenever possible. As *ex cathedra* central government policy, justified by reference to historical necessity and inherent ethical value, it was not likely to be anything else. A viable policy would have been based upon close attention to local variation and a respect for the incentives actually operating upon both immediate producers and producing organizations. Since both of these conditions were lacking, the agrarian policy of which the NMS was part ended up floating somewhat abstractly above the deep-textured reality of the North Vietnamese countryside, relevant primarily because the top leadership said so. Thus the political compromises inherent in the failed attempt to implement the NMS reflected opposing ideas and interests in the national polity.

Local Ideas and Interests in Historical Perspective

Local ideas and interests have to be seen historically, and they are illuminated by some comparisons between the cooperativized commune and its dynastic counterpart, the premodern commune. The striking points of similarity and difference center upon the local leadership's economic roles. Two major differences were the apparent fragmentation of corporate economic power in the cooperativized commune and the novel and extensive involvement in the direct control of production.[2]

Both systems were characterized by the existence of formal groupings of considerable authority at the level responsible for corporate liabilities to higher levels. And in both, a dual system of property rights reflected the partial freedom of local interests from higher-level interference.

It will be recalled that in the dynastic system the council of notables contained the powerful members of the commune and controlled the agent recognized by the state, the *ly truong*. The Commune was responsible to the state for taxes, corvée and military labor, and the maintenance of law and order. For full commune members the council recognized "citizenship," social ranking, religious ritual, and private property, while becoming the focus for political struggles. It was the source of communal land. The balance between communal and private land was determined largely by the extent to which the state's influence offset pressures upon members of the council to exploit communal land for family gain. Private land was seen as land subject to local interests and unregulated by the state, and in practice private land and communal land were farmed by similar methods.

The cooperativized system was a radical departure from this. On the face of it the cooperative's liability for tax payments and labor mobilization and its specified role as intermediary between cooperators and higher levels represent a striking continuity. The Party Committee in the commune (the *chi bo*), whose secretary was customarily chairman of the Supervisory Committee, appeared to provide a focal point for local authority comparable to the old council of notables. But the two were quite different, for a number of reasons. In the cooperativized commune there was no direct counterpart to the open and intense political struggles that had centered upon the "elections" to the council. Any intraparty conflicts remained comparatively exclusive. The changing pattern of alliances and social competition based upon status-ranking rituals, the giving of presents, and the related life-ceremonies (weddings, funerals, etc.) were conspicuously lacking. Only remnants remained in the feasts often given to visiting cadres, the high social status attached to education, and so forth. One possible reason for this was the apparent continuity of local leadership after the end of the "Correction of Errors" in many communes, and the need for the party's local leaders to become involved in all important functions in playing the leading role advocated by the party.

A second difference between the cooperativized commune and its dynastic forbear was the parallel party links with higher levels which in principle bypassed the cooperative structure. Although formally responsible for intermediary functions the Management Committee could not monopolize relations with superior levels as the council of notables had done, because the national party hierarchy allowed other elements in the commune to remove, for example, an established clique from the cooperative's management organs. The position was complicated by any use of overlapping personnel.

The production brigades appear to have played a key role in local affairs, and they performed important social functions quite outside their formal economic tasks. Their direct responsibility for collective production was a radical innovation that provided an apparently clear proof of the party's commitment to economic advance. Collective economic activity offered, in the control of production and output, a source of power that, in the general absence of comparable land concentrations, had no counterpart in the dynastic system. The terms of participation were quite different from anything previously encountered, with labor-sharing, the systems of points to determine distributional entitlement (at least in part), no formal possibility of job loss because of the "right-to-work," and an essentially geographical determination of the labor supply. Whatever the methods of surplus extraction the cooperator working in a brigade was not a wage-laborer. The extent and reality of collective production was basic to the distinction between nominal, active, and model cooperatives.

The dual property system (i.e., the coexistence of communal and private land) of the dynastic communes also had its formal counterpart in the cooperativized system. Cooperators retained formal private property rights to "orchards and ponds" as well as to their houses and household chattels. In practice, however, effective private property rights extended over the 5 percent land, which formally remained collective property. The balance between the claims of the collective and those of the cooperators partly depended upon the extent to which the local leadership was willing and able to follow prescribed policy. The precise nature of property relations within the cooperativized commune depended upon the effects of higher levels upon the local leadership; this was reminiscent of the traditional commune in which defense of communal land depended partly upon state intervention. But the deeper interactions between the collective and household sectors that resulted from collective production made the cooperativized commune a radical departure from earlier precedents. The important split between the inward-facing brigade leadership and the outward-facing Management Committee reflected the greater fragmentation of both formal and actual authority in the new system; the local Party Committee was not necessarily able to enforce unity around the prescriptions of the party center.

Thus although many aspects of the cooperativized commune recall its dynastic predecessor—most especially the importance of corporate structures and responsibilities to higher levels and the associated existence of a dual property system whose balance was affected by superior levels—the new system was a radical change. There was no longer any single institution capable of grouping in one body all those members of the commune holding either formal or actual power. The old social ranking and religious rituals had gone. Collective production was the most striking innovation in the new system, and it allowed for a far more sophisticated and complex interaction between the two property systems, where production was now carried out in quite different ways. But production technology remained largely unchanged, in the sense that a nineteenth-century peasant would have had little difficulty in comprehending the issues involved in, for example, the use of new seed varieties: water control, fertilizer, ploughing and harrowing, and so on. Apart from chemical fertilizer and machines there was little he or she could not have easily understood in the new ideas. Their "scientific" basis would naturally have been beyond comprehension; but the point at which such knowledge becomes relevant is debatable. A major problem in ensuring that modern ideas were disseminated in line with the party's commitment to economic advance was that the older generation—who often either could not or would not use better methods of cultivation—remained entrenched in positions of authority in the institutions of the cooperative and commune. This coexisted with the adverse immediate incentive structure and other reasons for opposing the introduction of new techniques. In such a social context formal positions of authority could be used as sources of personal or family gain.

Implications for the Likely Direction of Reform

The historical comparison brings out certain points that became increasingly important as policy began to change in 1979. First, while personal and family interests were likely to be paramount, strong arguments existed for maintaining institutions that could offset risk and limit excessive economic differentiation. Experience with the dynastic commune, the awful effects of land-loss during the colonial period, and common sense all gave good grounds for supporting some sort of corporate organization. Further, this could be used as a protective intermediary against the procurement demands of the new state if necessary. Second, the social position of those who held power in these bodies, although naturally strong, should be at minimum cost. Dynastic notables had become unpopular landlords during the French period. And Communist cadres could take some of the cooperative's rice for sale on the free market (e.g., in My Tho), discriminate in their allocation of duties and collective goods, and impose inefficient and top-heavy management systems in obedience to higher levels rather than their cooperators. If there were very many cadres, and they did not visibly contribute much to economic output, then their share of collective output could become a major burden. In a sense, then, one way of answering the Agrarian Question was to find a role for local cadres that allowed them to fulfill their political roles vis-à-vis both higher levels and the peasants while also minimizing their cost; if they were simultaneously efficient and technocratic positive contributors to net output, then all to the good, but if not then there should be as few of them as possible. The rural cadres were the base of the party-state hierarchy, and any change to the formal structure of rural social organization would have important implications for their positions.

High-level Theory and Hypernationalism

The obvious implications in previous chapters drawn from the detailed studies used to examine the sample must have been widely noted. The materials were, after all, mostly written by students at Hanoi University. Quotations from major speeches in chapter 4 showed that top leaders were fully aware of the policy unimplementability inherent in the Agrarian Question. The NMS was an attempt to deal with precisely that problem by strengthening cooperative management. The evidence suggests that official reports to the top leadership described events in the countryside reasonably accurately. The Red River Delta is not physically large, and members of their extended families should also have helped inform leaders about realities in the rural areas. Nonetheless, the validity of *Mac-Le-nin* (Marxism-Leninism) was a necessary part of the party's self-legitimization. As has been seen, in Vietnam this tended to produce idealistic ways of thinking and therefore a tendency to argue in terms of models. But although this goes some way toward explaining why and how the NMS came to be official policy, in many ways the precise reasons remain unknown.

The tendency toward the use of models in argument had severe adverse effects upon debate, leading to great difficulty in separating means from ends in the evaluation of policy implementation. This meant that many writers could not confront the basic issue of producer incentives and get at the real reasons why those in control of economic resources adopted certain strategies rather than others. This is perhaps inevitable when long-established official policy is widely unimplementable. The analysis of pig-rearing in chapter 8 showed that these strategies could in practice be quite sophisticated, relying upon a combination of the dual pricing system itself and different rights of disposal to secure economic resources from the cooperators.

A major problem in indigenous policy evaluation was this lack of clear distinction between means and ends. Policies were framed holistically, and perceived success depended as much upon the correctness of the methods used as upon the attainment of what other economists might naturally see as the end of policy. Great difficulties arose if a cooperative with an unorthodox management system could be criticized for that alone despite its success in other areas (e.g., efficiency of resource use or income growth rates). If ends cannot be clearly stated in isolation from the means for attaining them then the efficacy of existing institutions in meeting them cannot be properly discussed and appropriate action decided upon. It might be pointed out here though, that such behavior is almost inevitable when analysts are under great pressure from their environment and their superiors, and it is therefore likely whenever similar conditions arise. Vietnamese economists are neither more nor less stupid than those of other nationalities.

But it has also to be accepted that the idealism of wartime hypernationalism must have benefited popular mobilization. Propaganda required images and models, and these did not have to have the support of "scientific" investigation. Perhaps the lack of such a basis adds to their power. Here, though, it is important to recall the leadership's recommitment to the centralizing policies of the NMS in 1976, as they moved to reunify the country de jure by setting up the Socialist Republic and began to extend socialist property forms—including agricultural collectivization—into the Mekong Delta of South Vietnam. Thus the leadership continued with previous policies even after the need for mass popular mobilization was over and the country reunified. The six-year gap between reunification and the formal introduction of the "household contracting" system in 1981, coupled with the circumstances of its implementation, all suggest that the top leadership retained their commitment to socialism as they understood it, with its implicit belief in the moral authority of the center, based upon Marxism-Leninism and its implications for the historical necessity and ethical value of the Socialist Revolution in Vietnam.

This all implies that the national political compromises surrounding the Agrarian Question were at root to be found in the opposition between the different ideas and interests of the national leadership and groups at the rural periphery. The former were based upon Vietnamese perceptions of Marxism-Leninism. The latter were bound up with problems of corporate risk insurance, local power struggles, and the meaning and control of those local institutions that acted as protective intermediaries between peasants and local cadres on the one hand and the state on the other.

Economic Rationality and Economic Development

Criteria for evaluating economic policy should be sufficiently broad to accept widely differing values and goals. At a minimum, it is difficult to see how a policy can be rational if it does not possess:

1. Clearly identified goals that can be empirically measured, at least in outline;[3]

2. An appreciation of the nature of the incentives acting upon those in a position to allocate economic resources—with a knowledge of who those agents are;

3. A strong awareness of the uncertainty surrounding the prediction of economic behavior;[4]

4. A sense of the underlying notions of static and dynamic efficiency—a rational strategy is one that aims to attain current and future goals at least cost.[5]

A policy that stresses models (e.g., in the context of the NMS, Dinh Cong and Vu Thang) certainly possesses clearly identified goals—the models themselves. But these models are obviously not the sole goals, since higher incomes and accumulation are also important. The use of models in fact becomes a source of great confusion, and this came out clearly in the division between technocratic and organization and management viewpoints in chapter 10.

The nature of the sources used in this study has clearly had a deep effect upon its scope and therefore upon its conclusions. In using the observations and analyses of Vietnamese economists two key issues came continually to the fore: the explicit intention of the NMS to centralize control over economic activity, and the persistent adoption of various strategies by local interests to avoid such a shift in the local power balance. To the outsider, however, other issues—such as those discussed above—might be of equal or greater importance. For instance, the extent to which collective incomes were able to compensate peasant households for large differences in the distribution of labor resources and provide an insurance against the riskiness of the natural environment. Again, although it is reasonably clear that there existed a deep and complex pattern of social relations within the collectivity, an understanding of which was essential to obtaining a grasp of how economic activity was determined in practice, this was not a major concern of the sources. They can be, and have been, used to show the existence of such problems, and to help point to a better method of analysis. But they reflect, in their perspectives and concerns, some of the basic attitudes—if not social consciousness—of part of the North Vietnamese establishment. They are, after all, the writings of students from the premier university in North Vietnam.

From such considerations it is possible to draw certain conclusions. First, there was, during the period of this study, a powerful sense of the need to realize a certain organic social unity in the field of collective agricultural organization. People should be united and basically agreed about how things are done. Such conceptualizations naturally placed purely economic goals—such as yield maximization subject to distributional constraints—into a lower level of priority. They assert the validity of some degree of local political autonomy, and this is clearly consistent with Vietnamese Marxism-Leninism. The underlying sense of a need for unity has surely to be appreciated in the light of wartime experiences and the liberation struggle. Also, the long period of policy unimplementability suggested that if people did not agree it would be very difficult to force them to go along with official policy. Second, hard experience had showed the comparative impotency of these ideas in the face of local autonomy, local interests, and the various strategies employed to turn cooperatives into protective intermediaries. This study has emphasized the economic disincentives to participation in collective production. The sources, while confirming this, tended to stress the effects of local party disunity and/or the abilities of the Management Committee (chapter 10). Both tended to see the problem in terms of the pure effort of will needed for people to act in accordance with the party's conceptualization of social affairs and the meaning of the Socialist Revolution in the countryside. This suggests two things. First, that there would be pressure for new policies to reflect the above as well as more conventional issues presented at the beginning of the chapter. Second, that if one is to accept that social context has an effect upon people's perceptions and social consciousness, then this study should be taken, at least in part, as a subjective reflection of North Vietnamese views of their own society during a period of difficult postwar readjustment. Political control over printed matter and other channels of communication in itself provides insight into the outlook of those exercising that control.

Toward a New Compromise—Interests in a Developing System

One of the most striking characteristics of the NMS was the way in which it effectively inhibited growth and accumulation. Buoyant supplies of economic resources tended only to be available on the private plots, but accumulation based upon new inputs was meant to be reserved for the collective. Thus the nominal cooperative possessed an essentially stagnant economy, with very limited diversification and accumulation. In addition, static inefficiency was high, so that there was a potential for short-run output gains that active cooperative management teams were able to tap by "creatively modifying" the NMS. Since the party's commitment to national economic development commanded wide popular support this situation must have been almost universally unacceptable. Thus any form of accumulating system would be a viable alternative to the NMS. This almost naive point has important implications for the way in which policy shifted after 1979.

Within an accumulating system the power basis of local cadres can change rather dramatically, as increasing levels of output and production diversification create technocratic roles that give an economic justification for their social position. The NMS failed to do this because it did not create favorable expectations about the likely material consequences of its implementation. Furthermore, the higher income levels and output diversification of an accumulating system should usually help to reduce perceptions of risk.

From these considerations come certain basic predictions about the likely shape of a politically acceptable and implementable set of policy reforms.

First, some form of collective structure, labeled clearly as socialist, would persist. This would answer the top leaders' need for a validation of their long-held beliefs. This was important so long as they remained in power, without major political disturbances such as the period of the Gang of Four in China. Collective structures would also facilitate meeting perceived collective obligations to war invalids, the families of war-dead, and others. In poorer areas incapable of rapid growth it would also help offset the higher levels of risk associated with their lower income levels. Thus full-scale decollectivization would encounter opposition at both local and national levels.

Second, new policies would pay far greater attention to the direct material incentives operating on producers, and the overall static economic efficiency of the system. The daily experience of the absurd resource losses generated by the coexistence of inefficient collective production with the alternatives offered by cooperators' own-account activities ensured local support for such changes. The existence of considerable slack within the system, and the resulting potential for a sharp rise in output once incentives were improved, meant that national goals, especially of procurement and increased nonstaples supplies to industry, would also be met by such a policy.

Third, new policies would tend to seek ways of improving the economic effectiveness of local cadres. One solution was to find space for the rising technocratic local cadres in those areas of the collective economy where experience had shown their value—seeds, chemical fertilizer, and insecticides were clear examples. The value of placing other new inputs such as the machines used for ploughing and harrowing under such technocratic control is not so obvious. Coordinating functions, however, for instance in such areas as water control and production diversification strategies, offered considerable potential for such people. This shift in the social basis of local cadres would permit many of the educated younger people, previously frustrated by the conservatism of their elders, an outlet for their energies. A crucial element of this transition would probably be the role of the brigades and hamlets in supporting the social basis of the old local leaders.

Fourth, major ideological problems would arise in the attitude toward the likely expansion of household-based activities which, experience had shown, were in many cases economically more efficient than collective production. After years of stressing the virtues of collective work, and its socialist nature, private activities would have to be masked to some degree in order to avoid too much embarassment.

12

A Solution to the Agrarian Question?

This chapter looks at the way in which the policy thinking behind the NMS changed as the collectivized system came under great pressure in the late 1970s. As far as the organization of Northern collective agriculture was concerned, the most important element of the eventual response was the so-called output contract system (*khoan san pham*).

The Build-up to the Sixth Plenum of August 1979

From 1976 onward the existing difficulties with the collectivized system mounted steadily. In the North the macroeconomic tensions described in chapter 1 tended to increase rather than diminish. This suggested that basic lessons concerning the operation of the DRV's economic management system had not been learned from the experiences of the First Five-Year Plan (1961–65). Competition for resources between the socialist and "outside" economies in the North intensified, and adverse incentives increasingly diverted resources away from the low-price, administratively managed, inefficient state sectors. Effective legal constraints upon accumulation in nonsocialist sectors, coupled with the great inefficiency and waste within many socialist production units, severely inhibited the generation of domestic sources of accumulation. The economy began to settle into a low-level equilibrium where the state could only support itself in the face of resource loss to the outside economy through

194

its monopoly control over foreign aid supplies.[1]

A number of special factors heightened the systemic difficulties. Consumer goods from the South started to appear in the North once the border was opened, and this increased the demand for cash incomes with which to buy them. In addition, the state was forced to divert supplies to the South to prevent complete economic collapse after the military defeat in 1975. This reduced the volume of goods available to the state for exchange with the Northern peasantry at the same time as competition from the free market was increasing. As time passed, peacetime development of state industry and foreign aid proved insufficient to offset falling incentives to participation in collective production. Free-market food prices began to accelerate, indicating the extent of these underlying macroeconomic tensions.[2] The pattern of incentives shifted still further against rural collectives as the free market became more and more attractive to cooperators. Procurement in the North started to fall steeply,[3] and, as table II.7 shows, nonagricultural workers in the North were steadily forced to seek a higher proportion of their food requirements on the free market. As the decade came to an end, the macroeconomic factors behind cooperative nominalization and the Agrarian Question were becoming increasingly powerful.

The reunification of the country had in principle relaxed a basic constraint upon economic development by allowing the urban and industrial populations access to the Mekong Delta's large potential rice surplus. But efforts to collectivize the Mekong Delta rapidly began to encounter major difficulties, and after the 1978 drive against private sector traders and producers in the South deliveries to the state from the Mekong Delta slumped badly.[4]

In addition, there was very bad weather from 1977 until 1979, with particularly acute difficulties in 1978. The economic crisis was given a final further impetus by the Western and Chinese aid cuts of 1978–1980, which reduced absolutely the volume of goods in the hands of the state. Political uncertainties at the highest level rose in the wake of the Vietnamese Army's toppling of Kampuchea's Pol Pot regime in late 1978 and the Chinese "punitive incursion" into Vietnam's northern border provinces in early 1979.

By mid-1979 continued support for the NMS, which since the earliest days had had to avoid confronting the questions posed by the nominalization problem, necessitated almost intolerable obfuscation. Collapsing collective incomes were driving up grass-roots pressures to reduce collective production in favor of own-account activities. The long-term growth strategy was in tatters, as the reunified country had shown itself unable to create viable and domestically based accumulation processes well before the aid cuts gave a further twist to the downward spiral. The government had not proven its ability to manage peacetime construction. State industry had been able neither to break out of its chronic import dependency nor to avoid reliance upon food aid to feed its workers. And once Vietnam joined Comecon in 1979 and the Soviet Union became almost the sole source of aid, foreign pressures probably mounted for policy changes.[5]

As grass-roots pressure for further decollectivization mounted, the authorities initially responded by attempting direct regulation of incomes distribution in cooperatives to maintain procurement levels. The Second Plenum of mid-1977 confirmed the commitment to the Thai Binh Conference line of August 1974 by, inter alia, issuing a resolution that cooperatives were to set aside 10–15 percent of their land to grow feed for the collective pig herd.[6] The minister of agriculture, Vo Thuc Dong, was sacked in 1977 and replaced by the more senior Vo Chi Cong.

Chapter 4 mentioned Decree 75-CP of April 1974, which announced the abolition of the free market in staples. While the micro sources confirm beyond doubt that this was unimplementable, its underlying position on Northern cooperatives' procurement obligations was uncompromising:

> The distribution of staples inside agricultural cooperatives will still follow current regulations. For workers, the highest level of distribution must not exceed the food intake of a good worker in the same area. The remainder must be distributed in the form of cash calculated at the prices used for sales to the state. Distribution of staples inside the cooperative must still be according to work done. . . .
>
> If peasants still have surplus staples after they have fulfilled their obligations to the state, they must sell them to the state via the cooperative at incentive prices. Staples produced on the 5 percent land and not used for livestock must also be sold to the state at incentive prices. People who are supplied rice by the state but do not use it all must sell it back to the state. They are not allowed to give it away, sell it, or exchange it. (CB 15-4-74, p. 62)

> The state will not supply staples to any cooperator who stops work without permission or who leaves his cooperative to work illegally. (CB 15-4-74, p. 63)

The obvious intent behind these statements was to use administrative measures to prevent cooperators responding to immediate material incentives. The prohibition on free disposal of output from 5 percent land contradicted the Statute. A number of official documents confirm cooperatives' rising difficulties as collective incomes were squeezed between cooperators' diversion of resources into own-account activities and the state's pressure to maintain procurement levels.[7] These culminated in Decree 55-CP of March 1978, which appears to have induced some cooperative managements to abandon any attempt to base collective output distribution upon the workpoint system and revert to some rationing scheme. This decree was to figure in the debates about the reasons for the collapse of the NMS that occurred during 1980 (see below). There is little evidence, however, that during 1978 major policy changes were in the offing. Indeed, a decree early in the year tightened up on provincial rice purchases by preventing them from varying prices paid to producers.[8]

In 1979 matters came to a head, and many cooperatives began to show a sharp increase in the spontaneous adoption of various ways of extending cooperators' own-account activities and the noncollectively farmed area. Cooperative nominalization prior to then had probably permitted the real private-plot area to rise to perhaps 20 percent of cultivated land. Rapid inflation on the free market was continuously increasing the incentives for further incursions into collective land, and, as procurement levels came under further pressure the authorities faced a mounting crisis. But policy appears only to have started to move from the late summer. The Fifth Plenum of February 1979 had held to the old line, pushing for a further extension of the Vu Thang model in Thai Binh province of North Vietnam.[9]

In June a conference organized by the party daily *Nhan dan* in Thanh Hoa had examined the problems of pig-rearing in the wake of the 1977 decision to make all coperatives allocate 10–15 percent of their land to growing feed for them.[10] The conference, like the Fifth Plenum, showed no clear sign of what was to come. A frequent theme at the conference was the inherent difficulties involved in implementing a policy that stressed collective pig-rearing at a time when staples supplies were inadequate and free-market prices high. The basic premise, reiterated many times, was that higher staples output required more fertilizer, and that meant more pigs. Few asked openly how the initial fall in human consumption would be brought about. Many apparent successes had had to rely upon the households to rear pigs for the cooperative, which, as seen in chapter 8, usually meant exploiting the households' de facto abilities to sell on the free market. This flew in the face of the established official view of the advantages of collective pig herds. The following quotation from the conference shows the degree of obfuscation required of participants and the lack of overt criticism of existing policy:

> The common problem of the entire country, as in Thanh Hoa, is that if we wish to resolve the staples problem we must above all stimulate livestock, especially pig-rearing. The slow development of staples production has many reasons, among which is the shortage of fertilizer. The way to increase fertilizer supplies is through proper use of existing labor supplies and land. *That is our direction and way of working.*[11]

The Sixth Plenum and "Bung Ra"

In the face of these difficulties, and under great pressure, official policy finally started to unravel. By the end of the year, the central authorities had been forced to give quasi-legal justification to cooperators' desires to respond to immediate material incentives by at least two steps. These were, first, the sharp atmosphere change at the Sixth Plenum of August 1979 and, second, a small number of subsequent decrees of the Council of Ministers, the most important of which permitted cooperatives to "give out" idle land for their cooperators to farm directly. Over the winter of 1979–80 these changes facilitated a further sharp fall in the area of collective cultivation, and the process of nominalization began to become one of effective decollectivization.

It is hard to justify the idea that the Sixth Plenum introduced more than a profound change in atmosphere as far as agrarian policy was concerned. First, the meeting of the Central Committee only lasted one day.[12] Second, the plenum's published resolution did not actually mention agriculture at all, and it was in fact entitled "On the Direction and Tasks of Developing Consumer Goods Industry and Regional Industry." Its main focus was upon the problems facing state industry, which was suffering great difficulties in the wake of the Western and Chinese aid cuts. Here, it effectively opened the way to ending the state's monopoly of materials distribution through the system of direct administrative supply by encouraging industrial enterprises to search actively for inputs if they could not get them through the usual channels. This encouraged an expansion of strictly illegal horizontal relations between enterprises quite outside the plan that had already started in some areas. Local enterprises in particular began to expand direct links with agricultural producers to obtain inputs and food supplies. Official sanctioning of such behavior, which was greatly beneficial to the overall efficiency of industry, conclusively proves the reality of the change in political atmosphere that was occurring. Third, the relevant decrees of late 1979, as will be seen, did not deal with anything that could be called "fundamental change" in the collective system. Rather, they sought to carry out a holding action until some control could be established over an extremely unstable situation. This came in late 1980, with the output contract system. A new political compromise had to be reached.

As far as agriculture was concerned, the most important element of the Sixth Plenum Resolution was its recognition of the importance of material incentives and its implicit willingness to consider almost anything that would break the logjam and raise output by increasing the efficiency of resource utilization. The key phrase here was the need to encourage an "explosion" (*bung ra*) of output.[13] During the autumn two measures aided agricultural recovery in the North. First, September Decree 318-CP of the Council of Ministers dealt with the problem of "idle land."[14] Many cooperatives had been reporting falls in their cultivated areas as cooperators set about reducing the collectively farmed area. Cooperatives were now formally allowed to give out such "idle" land to be farmed noncollectively. In early October Decree 357-CP introduced incentives for peasants to rear livestock for meat:[15] rights of disposal were improved by allowing households to rear animals "without limit." The attempts to interfere in distribution within cooperatives continued, and in December the Government Council issued Decree 400-CP on income distribution within cooperatives. This stressed the need to distribute all of net collective staples output to cooperators as remuneration for work done, thus abolishing any tendency for rationing encouraged by Decree 55-CP mentioned above.[16] But in this document there is no mention of any system of contracting with cooperators.

These economic developments in the North, helped by the state's concessions to cooperators, rapidly boosted food supplies on the free market. As exchange-based relations continued to grow, the authorities started a series of measures to curb them that were to continue in coming years. Reports suggest that the output response in the North was extremely rapid,[17] and market curbs widely unwelcome. A decree of mid-October 1979 had reinforced existing regulations requiring traders to pay taxes and register, and to respect the state's monopoly of trade in various key commodities.[18] This revealed the limitations of the changes in the policy atmosphere brought about by the Sixth Plenum. Reforms to the internal organization of socialist economic units aiming to raise economic efficiency were tolerable; an expansion of spontaneous relations between such units outside the plan and, therefore (at least in principle), central control was a quite different matter, and many in the top leadership found such behavior far harder to accept. In the middle of 1980 the authorities reacted strongly to these unwelcome trends, introducing an important Politburo resolution on the "reform of policies for distribution and circulation (finance, prices, wages, trade, and the management of markets)."[19] These measures confirmed the boundaries beyond which party leaders were not yet prepared to go and heralded the establishment of a new policy toward the agricultural collectives.

Toward a New Political Compromise

Preventing a complete breakdown of the collectivized system appears to have been a major concern of policy makers through 1980. In meeting this goal, they essentially revised the basic thinking behind the NMS without fundamentally rejecting it. The key concept that provided continuity here was that of the "link" (khau*). By continuing to divide the labor process up theoretically in this way, and by simply allocating some of these steps to the cooperative and some to cooperators, the new policy created a continuing role for the cooperatives and therefore assisted recollectivization. But this time the greater economic efficiency entailed by prescribed policy meant that grass-roots pressures against the collectives were reduced. Once it had become clear that this was the authorities' new position, affairs could to some extent restabilize around the new compromise implied by this. The precise origins of this system are not absolutely clear because it was only one side of the story, and reliance has to be placed upon official sources in the absence of detailed grass-roots studies.[20]

The Economists' Debate

As the situation developed over the winter of 1979–80 it was accepted in some official quarters that a rural movement that "enthusiastically adopted the spirit of the Sixth Plenum" was in progress. Debate about the emerging "reformed system" began to surface during 1980. This involved a more or less explicit discussion of the reasons for the problems faced by the old methods (i.e., the NMS), which was of great interest. Many open participants in the debate were extremely self-critical, and the detrimental effects upon discussion of the long years of policy unimplementability are striking. A distinguished professor went into print with a remark effectively saying that social scientists had "stopped short at explaining party policy after it had been established and published"; instead, they should contribute to the determination of policy. Because reality was complex and its essence hard to grasp, there would necessarily be different interpretations of it.[21]

A leading defender of the principles behind the old policies was the academic Le Trong, whose article on Dinh Cong has already been cited. He published three pieces in NCKT during 1980. The first explicitly dealt with the use of "advanced models" in the dissemination of prescribed methods of agricultural organization. While acknowledging the influence of such problems as the excessive use of specialized brigades and the tendency for excessively large numbers of cadres to descend upon any cooperative that showed signs of advance, he defended the use of models and placed great emphasis upon the poor qualities of cadres and party members in explaining the many long-term "bad cases."[22] His prescriptions then mixed together the need to reinforce a "correct" organization of the party within the cooperative, coupled with an effective evaluation of the management cadres of the cooperative and more "economic" measures such as concentrating investment funds upon those areas giving the best results. These old themes suggesting the positive value of the NMS if properly implemented are quite consistent with the underlying philosophy behind the NMS, which, it will be recalled, sought to overcome the nominalization problem by strengthening cooperative managements.

He followed this piece with another long article criticizing egalitarian principles of income distribution in cooperatives and blaming such decrees as 55-CP for the collapse of collective labor incentives in the late 1970s. This gave a history of the methods utilized since 1959 that effectively attributed much of the difficulties of 1978–79 to the abandonment of distribution in kind.[23] Little of this in fact tallies with the information from the primary sources used elsewhere in this study, most especially the assertion that during 1974–77 official policy had induced Management Teams to rely upon a cash distribution of income while increasing staples supplies to the state. In addition, it ignores important arguments about the economic efficiency of different ways of organizing production. He then introduced the idea that the significance of the Sixth Plenum was that it had "abolished distribution by rations" and reaffirmed the need to distribute labor according to work done—that is, using a system of norms and workpoints. The main thrust of this article was the need for an operational and effective system of material incentives; there was no suggestion that an expansion of the household's economic responsibilities was either likely or advisable.

Le Trong's final contribution to the debate dealt directly with principles of internal labor organization in agricultural cooperatives. It also revealed one of the deepest and enduring premises behind the centralizing ethos of the NMS. The article was a clear defence of a reasonable application of the NMS in line with his previous writings (see the criticisms of the premature establishment of specialized brigades without any secure basis in production in chapter 4). It rested at root upon a simple but fundamental notion of the nature of the development processes, according to which the social division of labor necessarily increased during periods of social advance (such as, naturally, the Vietnamese Socialist Revolution).[24] This idea provided justification for the centralized and detailed control of labor of the NMS. The two basic flaws in this argument are those of periodization and logic. Even if it is accepted that historical periods of economic growth tend to be accompanied by various processes that do increase the social division of labor, this does not mean that one will necessarily generate growth by increasing the division of labor in one particular sector of the economy during a given period of time. It is quite simple to imagine situations where delaying such a shift would be beneficial to national output growth. In any case, the argument is too abstract, and therefore confusing: How does one unambiguously measure the "social division of labor"? If a cooperative increases output through higher efficiency generated by increasing the responsibility of cooperator households, while simultaneously setting up profitable teams selling insecticide and high-quality seeds to the farmers, then it has both reduced and increased the social division of labor. It is misleading to apply conclusions reached from analysis of long-term historical trends to medium-term economic problems.

Le Trong's position did not go unchallenged, and the reliance upon models came under attack. Another article in the same issue of NCKT presented sharp criticism of Dinh Cong. Nguyen Manh Huan (1980) asserted, inter alia, that the cooperative had received eight times the district average in financial assistance from the state—an effective critique. Its internal organization was also criticized, especially the lack of attention paid to the household economy.

The economists' open debate on such issues during 1980 brought into the open many of the frustrations felt by the participants.[25] It largely ended, however, with the publication in the early winter of Nguyen Huy's important article on forms of contracting. This emphasized the need to have (1) transitional forms, (2) a multiplicity of forms, and (3) a degree of choice (quyen lua chon) at the local level, above all by the collective peasantry. The article compared two basic forms of contract: first, the III-point contract between brigades and cooperative familiar from the NMS; second, a system of "product contracts with cooperators" reportedly devised by the town committee of Haiphong. This allocated to cooperators the four final links of rice production: caring for and lifting seedlings, transplanting seedlings, caring for the growing rice, and harvesting. Contracts were signed between brigade and cooperator households, and the collective was responsible for the other six links: ploughing and harrowing, seeds, supply of fertilizer, water, insecticides, and protection of the harvest. This system in principle allowed for unified management and created a cooperativewide division of labor;[26] he presented it as above all a way of organizing labor remuneration within the cooperative and a way of encouraging cooperators to work hard and effectively. This long article dealt with many other aspects of the situation but cannot be discussed fully here. It marked, however, a degree of formal public acceptance of the need both to improve material incentives and to retain a role for the collective in agricultural production. As such it represented the basis for a new political compromise.

A clear statement of the party's understanding of the situation and the new official policy also appeared in late 1980 with the publication of Huu Hanh's article on "Rice Contracts" in the party's theoretical journal, *Communist Studies*. This followed Nguyen Huy by stressing the need to view productive activity in rice cultivation as being divided into various links. Huu Hanh saw the cooperative as responsible for the following: ploughing and harrowing, seeds, arranging fertilizer supplies, water control, insecticides, and defending the harvest. The cooperative then allocated out the remaining links to its cooperators, using the familiar technical norms and the cooperative's plan, along with a system of penalties and bonuses payable in kind. This system was again attributed to Haiphong, and the author praised the city while criticizing Thanh Hoa province for delaying introduction of the new system. The latter province reportedly attempted to force a cooperative that had successfully introduced the *khoan* system in 1979 to abandon it and return "to the system of work contracts and penalty/bonuses in workpoints." This was a dismal failure, and the level of collective staples distribution apparently fell to 3 kg per capita per month—well below the target of 20 kg.[27]

Throughout this period information surfaced about the close connection between the *khoan* system and the de facto practices of the pre-1979 period. Prime Minister Pham van Dong's visit to a leading exponent of the system near Haiphong saw him praise the cooperative for having practiced it "covertly" (*chui*) since 1972.[28]

The Ninth Plenum of December 1980

The way was now clear for a formalization of the new policy. The key document here was apparently Order no. 100-CT/TW of the Party Secretariat issued on January 13, 1981.[29] A conference organized by the Ministry of Agriculture in Haiphong between January 3 and 7, 1981, saw the ministry present its report on the new system. The title of this report was revealingly entitled "Reinforce Agricultural Producer Cooperatives, Strengthen Contract Work."[30] The Ninth Plenum of December 1980 had given support at the highest level to the new policy, resolving that it was necessary to "expand the implementation and improvement of forms of contracting in agriculture" (p. 33).

The new policy presented the output contract system as one element of a number of measures aimed at improving cooperative management. Here criticism of the old system focused initially upon the premature amalgamation of cooperatives. The ministry reported that while many well-managed cooperatives had stayed intact during 1980, many of the weaker ones should now be broken up into smaller units. It gave detailed guidance as to how this was to be done (p. 41). The number of cultivation brigades had risen sharply—in the Northern deltas reportedly from an average of 126 workers and 44 hectares to 91 workers and 30 hectares (p. 49). No reason was given for this. The brigades were formally responsible for allocating output contracts to their members on the basis of the old III-point contracts they signed with the Management Committee. The ministry condemned specialized ploughing and harrowing brigades; irrigation teams were only appropriate if the water control system had been improved; fertilizer brigades should only be set up if they could directly help production (pp. 51–52).

The overall thrust of the new policy was therefore quite clearly toward a reintegration of the expanded own-account household activities into a new set of prescriptions regarding cooperative management. The output contracts were described as follows:

> [In addition, there are] "contracts for many links that are linked to final products, which are called for short "output contracts with cooperators" (*khoan san pham cho xa vien*). With this form, once the cooperative has its III-point contracts with brigades, the cooperative and the production brigade organize labor collectively at the level of the brigade or cooperative to carry out important or technically complicated work. A number of other links that are close to final output are contracted out to cooperators. The cooperator is responsible for delivering output to the brigade in accordance with the contract so that the brigade can deliver to the cooperative in accordance with the III-point contract. If the contracted amount is exceeded the cooperator receives a bonus; if there is a shortfall without good reason the cooperator must pay compensation . . . this form has a strong incentive effect upon the worker. (p. 56)

According to this report, output contracts had first been used for the winter crop (presumably that of 1979–80) and had then been extended by many cooperatives for rice and pigs. Haiphong city had issued a directive encouraging their use, as had Vinh Phu and Nghe Tinh provinces. Although they had had considerable success from the Xth month crop of 1980 onward, in most other provinces (Ha Bac, Ha Nam Ninh, Ha Son Binh, Hai Hung, and others) many cooperatives had started using them without any directives from higher levels. After the communique of the Party Secretariat they had become even more popular, with a majority of cooperatives in such provinces as Nghe Tinh, Thanh Hoa, and Ha Nam Ninh using them (p. 67). On average, "rice yields and output" (*sic*) had risen by 15–20 percent; the lowest rise was 5–10 percent while some areas had seen a jump of 30–40 percent with individual cooperatives doubling output and yields. These successes had made cooperators "more enthusiastic and closer to their collectives" (p. 68). But it was only after the Ninth Plenum that the party Secretariat had issued a formal directive on these contracts. This document (order no. 100-CT/TU) set five principles for cooperatives under the new system:

1. To manage closely, and use effectively, its means of production, above all the land, draught power, fertilizer, tools, and other material-technical bases of the collective.

2. To organize well the management and control of labor.

3. To have a plan, and to plan in accordance with the production zoning organized by the district, with ever-progressing economic-technical norms; all contracting units had to respect these regulations of the cooperative.

4. To keep direct control over output in order to ensure a proper distribution of it.

5. To develop the cooperative's right to autonomy and the cooperators' rights to collective mastery (pp. 60–61).

In addition, any cooperative that wanted to use output contracts had to meet a number of conditions. It had to have a clearly defined output development path that allowed it to meet its obligations toward the state; a proper system of economic norms and technical standards; secure supplies of the necessary inputs; clear regulations governing the operation of contracts signed between the Management Committee, the brigade head, and the cooperator; and close management based upon well-trained cadres, especially those acting as brigade heads (p. 63).

The important penalty-bonus system was not laid down in any great detail—"deciding what sort of bonus fund to use is entirely up to the cooperative in the light of concrete conditions" (p. 71).

By this time, however, major changes had also taken place in the regional allocation of resources for direct exchange between state and agricultural producers. Here the basic shift was toward a concentration of the limited available supplies upon the high potential rice surplus areas of the Mekong. As a consequence, policy had to stress regional autarchy in the poorer provinces.[31] Through 1980 and 1981, this helped secure a rapid recovery in procurement levels, mainly from the Mekong Delta provinces (see table 12.1). The basis for these policy shifts can be found in a number of decrees in the second half of 1980. At the same time, the authorities attempted to reduce farmers' perceptions of risk by fixing taxes and procurement quotas for a period of five years.[32] This meant that the causes of the immediate crisis had now passed, and some stability had begun to return to the overall economic situation.

Conclusions

The report of the Ministry of Agriculture clearly reveals the authorities' concern to protect the collectivized system. Its entire focus is upon the operation of the

Table 12.1

Staples Output and State Supplies (m. tonnes paddy equivalent)

		State supplies		
Year	Output	Domestic procurement	Imports	Total
1976	13.0	2.03	0.73	2.86
1977	12.9	1.84[a]	1.26[a]	3.10
1978	12.9	1.59[a]	1.47[a]	3.06
1979	13.7	1.40	1.77	3.17
1980	14.4	1.94	0.94	2.88
1981	15.1	2.52	0.45	2.97

Source: SLTK (1981), tables 32, 45, and 50. Data marked (a), SLTK (1978).

output contract *within* cooperatives. If the arguments in this book about the extent of cooperative nominalization are correct, this suggests that the output contracting system is perhaps best understood as an exercise in "recollectivization." This has important implications for the overall nature of the political compromises associated with the new policy. Compared with the NMS, the new policy had clearly highly beneficial effects upon producer incentives; much of the unwieldy apparatus of the NMS had been done away with, especially the specialized brigades. The overall attitude toward management methods within any given cooperative was now far from liberal, and there was no more talk of "models."

The wider implications for local power balances are not, however, immediately obvious. Furthermore, if the above arguments are indeed correct, the underlying reality of the *khoan san pham* system at this period is likely to be hard to establish. It is clear that, confronted with a severe economic crisis, the Vietnamese authorities had made certain concessions to grass-roots pressure for a reduction in the extent of the collective economy. The contracting system appeared to reestablish the household economy as the leading agent in the collectivized Northern agriculture. But here a number of important questions arise, and these are extremely hard to answer in the absence of proper fieldwork.

First, what was happening to the various channels for mobilizing surplus in Northern agriculture? Some reports suggested that contracting households were free to dispose of output as they wished, once they had met their contractual obligations to the cooperative, out of which procurement deliveries were still to be made. Taxes were indeed reportedly fixed for five years.[33] Here the repeated attempts by the government to control the free market show the underlying thrust of policy, which still sought to enforce the state's monopoly. Table 12.2 shows how this policy was unsuccessful, at least until 1981, for state employees continued to meet an increasing share of their food requirements on the free market. In

Table 12.2

**State Sector Supplies in Total Expenditure of
a Northern Worker Household (%)**

Year	Total	Food
1975	64.8	63.6
1976	66.5	65.7
1977	61.8	61.3
1978	58.4	55.3
1979	52.7	51.6
1980	42.6	34.6
1981	35.5	33.3

Source: SLTK (1981), table 59.

1980, indeed, the share supplied by the state fell by 17 percentage points. Nevertheless, it is important to establish just what control contracting households had in practice over disposal of the output of the fields they were allocated.

Second, how much control did contracting households have over what they chose to plant? Here the stress at national level upon the need for cooperatives to continue to plan production within them suggests that the authorities did not intend for them to play such a role.

Third, what was happening to the power base of the more conservative local cadres in the old monocultural brigades typical of the nominal cooperatives ? The new contracting system effectively ended the collective labor that they had been responsible for controlling. But it was not clear whether they were still responsible for appropriating output, or whether cooperators contracted directly with their cooperatives. The evidence here is contradictory, suggesting that the pattern was far from uniform.[34]

There are at least two alternative views of the output contracting system in North Vietnam. One is that it was in essence no more than an extension of piece-work remuneration. According to the formal system, contracting households were told what to produce and supplied with certain inputs so they could do so. They were allowed to retain a limited share of the resulting output to stimulate their efforts, but this share was set by the management of the cooperative. If it had previously operated extensive and illegal private plots in a highly nominalized cooperative, such a household would have lost much of its economic freedom. The cooperative would now determine what it produced, and its control over output disposal could also be much reduced. In compensation, new younger and better trained cadres would probably have forced out the "incompetent" old local leaders. The cooperative would probably be functioning more efficiently, and, given appropriate incentives from the state, accumulation could occur.

Higher efficiency would compensate for the effective recollectivization that the system had permitted in face of the widespread abandonment of the NMS.[35]

An alternative view is that the contract system greatly increased the economic independence and power of cooperator households. They gained an effective freedom to decide what they were to grow and how they were to dispose of it. Cooperatives were reduced to essentially little more than rent-takers, financing local communal services and delivering procurement while coordinating supply of technical services. This had adverse effects upon their ability to support valuable communal services such as child-care for working mothers, education, and so on. Furthermore, the increased autonomy of the household reinforced traditional and retrogressive patterns of paternalism.[36]

At present the research does not yet exist to allow one to answer these questions properly. There is likely to be substantial variation throughout the Red River Delta in the ways in which cooperatives operate. In the absence of much more work to establish what was actually happening at the grass-roots level, no firm conclusions can be drawn to decide between the various explanations. Certain points, however, do stand out.

First, staples output in the North appears to have increased sharply by 1981.[37] There is thus evidence for a once-and-for-all output gain in staples production derived from improved efficiency in static resource allocation. But thereafter, staples output reportedly stagnated. Second, official policy did not encourage development of household-based activities independent of the cooperative. There was no program of loans to peasant farmers, and the cooperative was still viewed as the point at which accumulation should occur. Third, the rigid thinking behind the NMS was replaced by a more flexible and pragmatic approach within which it was accepted that individual cooperatives would legitimately show a far greater degree of variation. Models were no longer featured as the main norm by which to judge cooperatives. On balance, then, a provisional judgment might be made that the changes that took place in 1979–80 represented a certain definite but limited advance on what had gone before.

The Meaning of the Vietnamese Revolution

Depending on the reader's point of view, a number of different conclusions can be drawn from this study. And for the convinced Communist or the long-time foreign supporter of the Vietnamese Revolution there are quite different problems.

The former expects difficulties and anticipates both the inevitability of errors and the need to overcome opposition in the certainty that the final outcome is historically determined: one day, communism will come. Thus the inappropriateness of the NMS is quite natural for such inexperienced comrades as those of the Vietnamese Workers' party, operating in an environment dominated by petty production and the sort of thinking that entails. The constant unwillingness of cooperators to work for "their" cooperatives is no indication of fundamental error, for the attraction they feel for own-account activities is simply the outcome of the presence of the ready potential for petty capitalism in such a country.

The latter, who seek to find in the surface harmony of North Vietnamese rural life confirmation of visions justifying their activity during the Vietnam War, surely face greater difficulties. Could peasants really have sent their children to fight while themselves going so far against the wishes and dictates of the party? Would the party really have continued with such policies in the face of popular opposition? But these are not questions that this study answers. To find out what North Vietnamese peasants really thought about the war and the party it would be necessary to ask them directly, in the proper circumstances. It is hoped that someday this will happen, if it has not already.

Cooperatives as a Solution to Problems of Rural Development

The experiences of North Vietnamese agriculture are usefully evaluated from a developmental perspective, so long as it is born in mind that the policy goals derived from it sharply differ from those of the Vietnamese Communist leadership. Anybody who takes the trouble to read enough of their writings will see this.

Agricultural producers in a climatic environment such as that of Vietnam's Northern deltas continue to face comparatively high risks. Now, given the generally low level of real incomes and the importance of securing basic subsistence, there is a convincing argument that an overriding goal of economic policy should have been the maintenance of some balance between the growth of staples production and the (largely exogenous) increase in population. The lack of stable export earnings to finance food imports strengthens this argument. Between 1960–65 and 1973–75 the region appears to have shifted from an approximate self-sufficiency in food to a substantial dependence upon imports. While any overall judgment on policies of self-sufficiency in food in developing countries remains contentious, the failure of North Vietnamese agriculture to prevent this shift to import dependency represents a considerable shortcoming of the cooperativization policy.

One basic explanation for low agricultural output growth during the period is that the adapted system of cooperativized agriculture (described here as the nominal cooperative) constrained technical innovation and direct accumulation. The "protective intermediary" syndrome could be reinforced by the continual attempts to strengthen the cooperatives, and especially by accelerated amalgamation. For this and other reasons the critical problems faced in the late 1970s cannot be isolated from previous policy decisions. The nominal cooperative was a product, at least in part, of the attitude taken toward the "correct" method of management of cooperatives as well as of the macroeconomic environment. Rather than helping the situation, the NMS tended to make matters worse.

Although a strict quantification of such aggregate variables as real cooperator incomes and growth rates of real consumption is impossible, issues relating to the effects of cooperativization upon cooperator well-being cannot be viewed ahistorically. Anyone who has studied the effects of the colonial regime upon the Vietnamese peasantry cannot but be impressed by the advances made under the new system during the period under study. Basic income support, social and medical services, an absence of systematic violence, abolition of extreme disparities in standards of living and consumption, basic education, and a certain stability in everyday life must surely be placed high upon any scale of values. But beside these factors must be placed the probability that basic subsistence was often not guaranteed by collective incomes, while access to extra land was determined by relations with the relevant section of the management team. Certain groups clearly gained from privileged access to economic resources (e.g., some dominant local cadres, free marketeers, those with pull in higher levels of the party or state apparatus). While on any simple judgment real incomes and income distribution had improved from the colonial period, ample grounds for criticism of the existing state of affairs could easily be found. The abandonment of the NMS in 1979–80 suggests that there always had been better alternative policies.

In the absence of detailed field surveys the evaluation of noneconomic factors is even more difficult. The peasants' attitudes toward the various types of cooperative cannot really be judged. For instance, party hegemony over political and intellectual life and restraints upon religious activity may or may not have led to the same responses among Vietnamese peasants as they might for Western intellectuals. Again, while the effects of the system on the roles played by women may easily be felt to be inadequate, in that men apparently continued to dominate affairs, the more liberal attitudes to marriage and intrafamily relations (such as family planning) may have generated strong positive feelings. Because of the wide role played by the cooperative and its subunits in rural life such things were probably extremely important influences on peasants' overall attitude to the cooperative movement. Furthermore, the "undemocratic" nature of a cooperative's management might not have been much criticized if it "delivered the goods." On the other hand, however, peasants may have felt that in practice the system was so far from that promised that it was fundamentally corrupt and hypocritical. In the present state of knowledge, both views are little more than conjectures.

A comparison of cooperativization with alternative policies inevitably requires a subjective, counterfactual approach. Yet it is not easy to imagine a conventional Marxist-Leninist party in the late 1950s as dependent upon Soviet and Chinese aid as was the Vietnamese not collectivizing agriculture. It should be remembered that collectivization in Vietnam occurred after Chinese debates on the issue had been concluded in favor of agricultural collectivization. It is more appropriate, perhaps, to consider alternative policies to those adopted after the initial movement to join cooperatives was completed. Here the rationale for the continual attempts to reinforce the cooperative by extending its prescribed area of operations (e.g., the shift to higher-order cooperatives, the policy of amalgamations and the NMS itself) can well be criticized. Yet this followed Comecon/ CMEA orthodoxy. It might even be argued that the NMS and earlier measures to improve cooperative management were essentially no more than responses to policy nonimplementation, and, despite the repeated failures, simply more of the same.

One major difficulty with the NMS was the frequent absence of any economic basis for integrated production at the level of the cooperative. Output contracting allowed management teams to concentrate upon those activities where the cooperative could offer services valued highly enough to overcome resistance to technical innovation, such as seeds provision. If these services provided sufficient benefits, then that should have supplied adequate stimulus for cooperator participation, and it would then have been unnecessary to attempt the wholesale shift in the local power balance entailed by the NMS.

An important element of the NMS, which does not really come out from the developmental approach of the study, is the way in which it generated jobs and activities for the new generation of educated younger cadres who were starting to come out of the various state training establishments in increasing numbers from the mid–1960s. With a strongly interventionist set of assumptions about their proper role in society, they found in the NMS an apparent answer to their need for employment. With nearly fifty cadres required to run the system in an average-sized commune, at least 250,000 jobs were available in the North alone. The political role played by such a group and its interaction with the old leaders at the local and national levels surfaced in their complaints about conservatism, and there were well-documented conflicts between the younger, better-educated cadres and the conservative old leaders in the communes. For such people the output contracting system provided a major advance.

Do Cooperatives Work?

This has not been a comparative study, and Vietnamese experience with cooperativized agriculture has not been placed beside that of other countries. Certain problems encountered in the attempt to analyze socioeconomic behavior in the case of North Vietnam may, however, have more general applicability. Two in particular come to mind. First, the central role of conflicts to control economic resources. Basic issues that arose time and time again concerned conflicts between various identifiable groups over output distribution, control over the production process, and access to means of production (including labor). No attempt has been made to describe these groups (usually identified with reference to the cooperative structure and levels of management) as "classes," but systematic cleavages within the cooperative's management team could be and were traced (chapter 3) to different positions in these conflicts. The brigade leadership's proximity to the labor process in collective activity and the Management Committee's distributive functions in relation with superior levels provide two examples. From the frequent reference to poor labor discipline and inadequate direct accumulation, and to the conflicts between cadres and the household sector about the livestock branch, any idea that cooperators were in control of "their" cooperative would obviously be naive: access to resources was dependent upon social position, that is, the institutions of the cooperative, the party, wider kinship groups, personal contacts, and alliances. Any attempt to analyze cooperativized agriculture should take explicit account of all these issues (circumscribed in this study by the all-embracing concept of the collectivity).

Second, the potentially dynamic role of the cooperative. Under certain (albeit unrealistic) conditions, some basic arguments for cooperativization were valid: the concentration of resources could lead both to direct accumulation and to technical improvement. The centralization of economic power implied by collectivization did not necessarily result in the use of that power in a "progressive" manner, as prescribed by policy makers. In some instances the outcome appears to have been quite the opposite. But despite this, cooperatives could still be made to function as comparatively efficient users of existing resources, and even serve as a means for increased resource mobilization: Van Tien was an obvious example of a cooperative not in receipt of the favorable input levels seen in models. While the cooperative form was not, therefore, a sufficient condition for advance, it is impossible to judge whether it was necessary, for the sources dealt only with the cooperativized system. Such an ahistorical argument has little value so long as the party continues to present cooperatives as an immutable historical necessity.

Appendix I

The DRV Food Balance

Table I.1

Staples Availability from Domestic Production

Year	Crude estimate of: Population (mil.)	Crude estimate of: Staples production (mil. tonnes)	Per capita per month availability, kg: A) Unmilled	Per capita per month availability, kg: B) Milled
1960	16.10	4.698	24.3	15.8
1961	16.57	5.201	26.2	17.0
1962	17.06	5.173	25.3	16.4
1963	17.57	5.013	23.8	15.5
1964	18.09	5.515	24.4	15.9
1965	18.63; 18.27*	5.562	24.9–25.4	16.2–16.5
1966	19.2	5.100	22.1	14.1
1967	19.78	5.398	22.7	14.7
1968	20.38	4.629	18.9	12.3
1969	21.00	4.709	18.7	12.1
1970	21.84	5.279	20.3	13.2
1971	22.30; 21.15*	4.921	18.4–19.4	12.0–12.6
1972	22.66; 21.48*	5.742	21.2–22.4	13.8–14.6
1973	23.14; 21.93*	5.190	18.7–19.7	12.1–12.8
1974	23.95; 22.70*	6.277	21.8–23.0	14.2–14.9
1975	24.55; 23.35*	5.491	18.6–19.6	12.1–12.7
Averages				
1960–65	17.4	5.2	25	16
1965–68	19.8	5.0	21	14
1969–72	21.9	5.2	20	13
1973–75	23.9–22.7	5.6	19.1–20.6	12.8–13.4

Sources: KTVH, tables 8 and 78, and Hanoi II (for population data marked *).

Notes: Population data from intercensus years calculated by assuming constant percentage growth; aggregation of nonrice staples to give the total in KTVH, table 78, is understood to have used ratios of 1:1 by weight for maize and 3:1 for potatoes (Hanoi I); the ratio used for the crude adjustment to milled rice basis from the paddy equivalent measure is 0.65 (1 kg of paddy was usually assumed for statistical purposes to give 0.70 kg of milled rice [Hanoi I]; 5 percent has been added for seeds, etc.). To the extent that other losses existed this would exacerbate the tendency of the data to overestimate actual supplies available to feed the population.

Table I.2

Soviet Food Exports to the DRV (tonnes)

Year	Wheat	Wheat flour	Rice
1958	—	—	—
1959	—	—	—
1960	—	—	—
1961	—	20,100*	—
1962	—	9,000	—
1963	—	2,700	—
1964	—	1,500	—
1965	—	1,500	—
1966	—	10,700	—
1967	—	39,100	—
1968	—	241,700	—
1969	1,975	225,000	—
1970	4,128	426,100	—
1971	—	301,300	—
1972	—	1,300	29,400
1973	—	251,265	52,809
1974	—	623,764	47,823
1975	125,000	201,058	—
1976	200,000	207,860	—

Source: *Foreign Trade of Soviet Union—Statistical Yearbook,* various years ("Wheat"—entry no. 70001; "Wheat flour"—entry no. 82001; "Rice"—entry no. 8200307).

*"All meal, flour."

Appendix II

Details of DRV Agricultural Production

To remove some of the distortions to the data caused by year-to-year climatic changes, and to give a simple idea of longer-term trends, the data have been averaged over four periods as follows:

1960–65: The six years of peaceful "socialist construction" including the last year of the Three-Year Plan (1958–1960) and the First Five-Year Plan.

1966–68. The three years of the first intensive U.S. bombings, culminating in the crisis year of 1968.

1969–72: The four years of war before the Paris Agreements, with intermittent bombing.

1975–75: The three relatively peaceful years before the fall of the Saigon government in 1975.

Table II.1

Rice Production

	1960–65	1966–68	1969–72	1973–75
Spring and Vth month				
Sown area, thous. ha	919	908	921	967
Yield, paddy tonnes/ha	1.74	1.74	2.12	2.36
Production, mil. t	1.60	1.58	1.95	2.28
Autumn and Xth month				
Sown area, thous. ha	1,457	1,310	1,235	1,235
Yield, paddy tonnes/ha	1.88	1.88	1.94	2.13
Production, mil. t	2.74	2.46	2.40	2.63
Total				
Sown area, mil. ha	2.38	2.22	2.16	2.20
Yield, paddy tonnes/ha	1.82	1.82	2.01	2.23
Production, mil. t	4.34	4.04	4.35	4.91

Source: KTVH, tables 79, 80, 81.

Note: Yield averages are calculated from the acreage and production averages.

Table II.2

Secondary Staples Production

	1960–65	1966–68	1969–72	1973–75
Potatoes				
Sown area, thous. ha	187	243	187	148
Yield, tonnes/ha	4.02	5.37	4.98	4.81
Production, thous. t	752	1,305	931	711
Maize				
Sown area, thous. ha	229	228	204	204
Yield, tonnes/ha	1.10	1.00	1.03	1.04
Production, thous. t.	252	227	209	211
Manioc				
Sown area, thous. ha	98	114	90	83
Yield, tonnes/ha	7.47	6.99	7.46	8.21
Production, thous. t.	732	797	671	682
Water potatoes and taro				
Sown area, thous. ha	—	17	12	10
Yield, tonnes/ha	—	9.88	11.4	11.2
Production, thous. t	—	168	137	112
Total (paddy equivalent)	0.85	1.00	0.81	0.74

Sources: KTVH, tables 82, 83, 84.

Note: Yield averages are calculated from the acreage and production average; the figure for "total" nonrice staples includes various other minor items.

Table II.3

Livestock (thousands of head)

	1960–65	1966–68	1969–72	1973–75
Water buffalo	1,510	1,640	1,700	1,770
Cattle	790	750	700	650
Total	2,300	2,390	2,400	2,420
Pigs (>2 months)	4,310	5,120	5,510	6,440
% in collective pig herds	(1965) 5.7%			(1973) 11.2% (1975) 9.7%
Horses	48	70	104	122
Goats	75	118	135	140
Poultry (mil.)	24.8	28.0	30.5	34.8

Sources: KTVH, table 92, and, for pigs in collective pig herd, Hanoi II.

Table II.4

Water Control in the DRV

	1965	1974	1975
Pumps	4,800	23,260	23,700
of which cooperative property	2,460	13,950	16,630

Water-works Constructed by the State to End of 1975
Thousand of Hectares of Land Served

	Capacity			Actual productivity	
	Drainage	Irrigation		Drainage	Irrigation
Large-scale works					
Lakes (11)	—	48		—	20 (40%)
Dikes (11)	36	164		30 (83%)	111 (68%)
Ditches (19)	261	246		65 (25%)	197 (80%)
Pumping stations (54)	117	307		106 (91%)	224 (73%)
Independent works					
Lakes (228)	4	65		2 (50%)	33 (51%)
Dikes (221)	—	34		—	27 (79%)
Ditches (187)	45	94		25 (55%)	33 (35%)
Pumping stations (292)	82		93	49 (60%)	59 (63%)
Pumping stations outside the unified system (239)	—	29		—	23 (79%)

Sources: Hanoi II, KTVH, tables 46 and 75. There is no information on pump sizes.

Table II.5

Machinery and Electricity Availability

	1965	1975
Machines available per 1,000 workers[a] in cooperatives		
Motors	0.5	4.5
Pumps	1.0	4.5
Standard tractors[b] in use	0.2	2.6
Threshers	—	1.5
Mills	0.5	1.5
% of sown acreage machine-ploughed	2%	10%
Electricity supplied per 1,000 workers[a] in cooperatives, million kW-hours	0.01	0.03

Source: Hanoi II.

[a]Number of agricultural workers of working age (i.e., women 16–55; men 16–60 [Hanoi I].
[b]The standard tractor is one of 15 hp (Hanoi II). Data probably include machine belonging to the state.
Electricity data may exclude power generated by cooperative-owned generators.

Table II.6

Per Capita Income in an Agricultural Cooperator Family

		Of which:		
Year	Total	Income from cooperative	Income from "minor household economy"	Other income
1976	100.9	100.4 (34.6%)	101.2 (54.5%)	100.9 (10.9%)
1977	109.6	91.6 (29.1%)	123.5 (61.2%)	97.5 (9.7%)
1978	110.4	92.7 (29.2%)	124.0 (61.0%)	99.5 (9.8%)
1979	132.7	93.0 (24.4%)	164.0 (67.1%)	103.4 (8.5%)
1980	126.6	103.9 (28.5%)	158.0 (61.1%)	80.9 (10.4%)
1981	185.9	127.1 (23.8%)	222.2 (64.9%)	192.1 (11.3%)

Source: SLTK (1981), table 60.

Table II.7

Retail Sales

Year	Total	Of which: Organized market	Unorganized market
1976	12.0	5.8	6.2
1977	13.8	6.6 (+14.5%)	7.2 (+16.3%)
1978	15.3	7.6 (+14.1%)	7.7 (+ 7.1%)
1979	16.7	8.3 (+10.1%)	8.4 (+ 8.2%)
1980	22.7	8.4 (+0.6%)	14.3 (+70.6%)
1981	44.4	19.0 (+126.4%)	.5 (+78.3%)

Source: SLTK (1981), table 43.

Notes

Preface

1. This book is essentially a revised and updated version of the dissertation. The main changes are the omission of a number of tables, considerable textual revision, and the inclusion of some new material.

2. See Bray (1983) for a valuable discussion of the implications of the technical differences between wet-rice cultivation and grain farming in Western Europe.

3. "Boi vi su hieu biet la co han va su chua hieu biet la vo han" (Because Understanding Is Limited and Misunderstanding Limitless), Hoang Kim Giao, "Ve viec van dung quy luat quan he san xuat phu hop voi tinh chat va trinh do phat trien cua luc luong san xuat trong quan ly kinh te nuoc ta o thoi ky qua do" (On the Use of the Law of Correspondence Between Production Relations and the Quality and Level of Development of Productive Forces in Economic Management During the Transition Period in Our Country), NCKT 5 (1985, 11).

4. Conference organized by the Social Science Research Council, The Hague, 1982.

Chapter 1

1. Davies (1980a); Lewin (1975, 97–157).

2. Tuan and Crook (1983); Donnithorne (1967, chaps. 2, 3).

3. Swain (1981); Wadekin (1982).

4. Neither Le Chau (1966) nor Leon Lavallee (1971) provides any substantial analysis of the macroeconomic problems discussed below, which are essential to any understanding of the domestic origins of policy changes. Spoor (1985) goes a long way toward closing this gap; see also Vickerman (1984) for an efficient and valuable use of the abundant translations available. Judith Appleton's bibliography in White (1982) provides a list of English- and French-language materials on Vietnamese development published in Vietnam.

5. Fforde (1981; 1983) surveys existing Vietnamese and French materials on the history of rural society from the nineteenth century to cooperativization in the late 1950s. The macroeconomic evolution of the DRV is discussed in Fforde and Paine (1987) and Fforde (1982, chapter 2). Fforde and Paine contains all the tables from the Vietnamese collection referred to here as KTVH.

6. KTVH, table 69. In 1975, 90 percent of cooperatives were of the so-called higher type—see below. Here as elsewhere the data refer to all peasant households and therefore include nonethnic Vietnamese (i.e., non-*kinh*) minorities.

7. Wadekin (1982, 105).

8. Truong Nhu Tang (1986, 248).

9. But see Fforde (1985) for a look at Soviet-Vietnamese relations.

10. C.f. Government Council Decree 61-CP of May 4, 1976.

11. For an early statement on the *khoan* system, and pointed comments upon its nonadoption in some quarters, see Huu Hanh (1980).

12. The standard Western work on nineteenth-century Vietnamese government and administration remains Woodside (1971). The assertion that the Tay-Son "reunited" Vietnam is highly contentious. This is partly the result of the natural desire on the part of North Vietnamese historians to deny credit for reunification to the Nguyen dynasty (1802–1945). Hodgkin (1981) presents arguments for accepting North Vietnamese historiography. For a further discussion see Woodside (1971, 2–4). It should be noted that the Tay Son never really succeeded in having both Saigon and Hanoi under a single administration.

13. Woodside (1971, 158–68) on the *bao giap* controversy of the 1800s. The discussion here refers throughout to the traditional concentrated settlements of the North and Center and not to the South (Cochinchina), where, in a frontier area, other patterns prevailed.

14. Baker (1979). This section draws upon Fforde (1981; 1983), which was summarized in Fforde (1982, chapter 1).

15. These were hardly "elections" in the accepted modern sense of the word. "The old, the retired mandarins, degree-holders, intellectuals, ex-mayors, ex-village officials etc. . . . made up the Council" (Phan Khoang 1966, 42); see also Phan Ke Binh (1975, 159–63).

16. Chapter 11 discusses the structural differences between the cooperativized system and the dynastic commune. It should be stressed that many aspects of the traditional commune are not well understood. Of great importance to comprehending the relative lack of land concentration in the precolonial period must be the notables' attitudes toward excessive private landholdings. The hostility of the court to such interferences with the local body politic has been well documented. The attitude of local interests other than the notables was also crucial to the relative lack of land concentration in the precolonial period, however, and this factor is also inadequately researched. Recent work has further complicated matters by suggesting a redefinition of our understanding of the nature of "communal" and "private" land. Nguyen Huu Nghinh and Bui Huy Lo (1978) suggest that "private" land was seen as land that was subject to communal control, in distinction to "communal" land, which was regulated by the state. They quote examples of the communalization of private land to support this. This means that private property in land should certainly not be seen as something close to Western notions involving largely unhindered personal rights of use and disposal. Instead, the picture is more one of a dual pattern of rights, both largely administered by the commune, that reflected the state's rights to levy taxes and so forth as well as those of the commune itself.

17. Scott (1976) presents the classic "moral economy" analysis of these events, rightly stressing the value attached by peasants to social institutions that could help them avoid risk—that is, the consequences of harvest failures. Huynh Kim Khanh (1982, 167–70) gives an extremely interesting interpretation of Ho Chi Minh's possible involvement in the risings.

18. Tran Phuong (1968, table 7), largely reproduced in Moise (1977, 260). It is also true, if these figures are to be believed, that more land was redistributed before the Land Reform than during it even if communal land is excluded. These statistics, derived from the official history of the period, suggest that the relative emphasis of both White (1981) and Moise (1977) on the Land Reform campaign proper is slightly—if unintendedly—deceptive. If one is interested in the overall pattern of change in rural power structures between 1945 and the collectivization campaign of 1959–1960, such evidence suggests that great attention should be paid to developments before the beginnings of Land Reform proper around 1953. This might help explain the relatively extreme reaction of the peasantry to Land Reform and the resulting need for the Correction of Errors campaign. It could also contribute to a better understanding of local power structures after collectivization and the nominalization process described below.

19. Moise (1977, 254–55).

20. The 1980 Constitution of the unified Socialist Republic of Vietnam reversed this, declaring that all land belonged to the state.

21. Moise (1977, 403); Tran Phuong (1968, 188–217).

22. KTVH, table 69. State farms were of peripheral importance in delta agriculture and were concentrated in the plantation regions, usually in the uplands. Here as elsewhere the data refer to all peasant households and therefore include nonethnic Vietnamese minorities.

23. Ibid.

24. Gordon asserts that those outside cooperatives in 1960 were "mainly rich or upper-middle peasants" (1981, 41). This ignores the need to provide some critical assessment of the piecemeal data upon which it rests. See Vickerman (1982) for a full discussion. If the role of corporate economic power in Vietnamese rural society was as important as is argued in this study, then the analytical meaning of "class struggle" becomes somewhat unclear. This is because concepts of class and class struggle are commonly tied to a concept of property rights within which there is some close and stable relationship between the legally defined agency of ownership and some group or individual. Thus a landowning class is made up of those individuals who own land; both as a group and as individuals they stand in opposition to those who do not own land. But when ownership is collective, this opposition is unclear. For example, if a landowner joins a group, his landholdings and associated social position contribute to that of the group; if, on the other hand, somebody becomes a member of a group holding collective property rights—for whatever reason—then they gain correspondingly so long as they remain a member. Thus economic power derives from the collective and not from the individual or household. The concept of "class struggle" bases itself on the idea that economic power derives from the individual as a member of a class. This is not true when economic power comes from a collective. How do you struggle against "exploitative classes" when those belonging to the collective's governing institutions may simply resign en masse without changing at all the rules of the game, for the new rulers will have the same powers as the old?

25. Informal conversation with Vietnamese peasant, Red River Delta, 1979. If instead they were to bring their land into the cooperative, part of the output from it would have been kept back by the cooperative to pay those who worked it and meet other costs—the perceived "labor charge." Thus by keeping it as fully private land the family could retain control over more output.

26. This position draws heavily on the work of Kalecki (1972) and Kornai (1980).

27. This conclusion is fairly certain. By the mid-1970s free-market prices were of the order of ten times state purchasing prices. At such levels 20 kg of paddy, an amount easily carried, would be worth more than a month's wages for a state employee. This alone suggests that the balance of incentives in the mid-1970s was strongly against the collective sector, especially at the margin. The balance of incentives appears to have moved particu-

larly sharply against the collective economy in two periods. First, during the later stages of the First Five-Year Plan (1961–65), as the pressure of the industrialization program pushed up free-market prices. Second, in the late 1960s, before the adverse effects of heavy U.S. bombing and poor harvests, coupled with continual high procurement levels, were offset by large food imports.

28. KTVH, table 37.
29. KTVH, tables 8, 78, 112.
30. KTVH, table 134.
31. To this must be added the frequent problem of below-subsistence levels of collective staples distribution.
32. Phan Khanh (1981).
33. KTVH, table 6; Naval Intelligence Division (1943, 64–72).
34. Dao The An (1973); *Vietnam Courier* (April 1978).
35. Conversation with manager of Yen So cooperative, outskirts of Hanoi, 1979.
36. KTVH, table 72.
37. Ngo Vinh Long (1973).
38. Data here and below summarize those in the appendix 1.
39. The possible origins of this decline are discussed further in chapter 4.
40. KTVH, table 46.
41. KTVH, tables 70, 73.
42. KTVH, table 8.
43. Le Trong (1980, 26).
44. Price index numbers published in the *Course in Agricultural Economics* (n.d.) suggest that state purchasing prices of agricultural goods rose over 45 percent over the period 1965–1975 while state selling prices were either constant or falling (p. 72). It is of course movement in these prices relative to the relevant alternatives that is important.

Chapter 2

1. KTVH, table 73; the data refer to labor in the state sector of agriculture. The data on the rural population come from KTVH, table 8.
2. In practice the cooperative and the commune were institutions grouping powerful inhabitants whose informal activities could have a dominant affect on community life. But in principle economic functions were the responsibility of the cooperative.
3. The above discussion is necessarily abbreviated and should not be taken as more than an indication of the general pattern. For instance, in some areas of Nghe Tinh province in Central Vietnam communes do not have the two internal levels of village and hamlet; also, in Ha Son Binh province there is an independent village without a commune near Ha Dong town (Van Phuc)—personal communication.
4. The accepted Vietnamese for cooperator is *xa vien*, and for cooperative, *hop tac xa*. See the "Note on Sino-Vietnamese Terms" in the glossary for a systematic comparison with Chinese terminology.
5. The reformers often found themselves in only nominal control of a commune-level cooperative, where real authority centered upon hamlet-based brigades.
6. The basis for the following paragraph is Hanoi I.
7. From a conversation in 1979 with the accountant of a model cooperative, around 30 km north of Hanoi, I learned that model cooperatives were favorably treated but were still limited to under 1,000 dong cash on hand.
8. A frequent term of mild abuse for certain people in circumstances that may be imagined was "district cadre" (*can bo huyen*)—personal observation, Hanoi, 1979.
9. This structure had clear parallels with the "parallel hierarchies" of the Resistance period (*Revue Militaire d'information* 281 [February/November 1957]).

10. A contributor to the debates about the NMS during the early 1980s put the likely number of cooperative cadres as far higher—100 (The Dat 1981, 230).

11. Hanoi I.

12. If the main source of household income is not the cooperative, and the correct focus for analysis of collective activity is the brigade, then one reason why the term "collective property" is misleading is that the formal structure then provides a forum for conflict rather than a guide to the interests involved.

13. The so-called higher-level cooperative was one where there was, in principle, no remuneration for property and distribution was primarily (around 70 percent of total income: Hanoi I) dependent upon workpoints earned. By the mid-1960s it had largely replaced the "lower-level" cooperative where property holders still received rents, and by 1975 only 2.5 percent of peasants were in "lower" cooperatives, probably predominantly outside the deltas (KTVH, table 69).

14. The most complete is "Dieu le hop tac xa san xuat nong nghiep bac cao" (type-written, photocopy in my possession). Obtained from the University of Hanoi in 1979 when another was "in preparation," the copy is apparently from a library and has been annotated. The document was said to be of contemporary validity. It is interesting that the Statute appears never to have been formally approved by any legal body. This version, with an introduction, is translated in Fforde (1984). The other two versions are both summaries translated into English: "Digest of the Statute on High-level Agricultural Cooperatives," U.S. Mission in Vietnam, Saigon (October 1970), and "Statute of Agricultural Coopera-tives (High Level)," Rural Publishing House unofficial draft translation (September 1974). Of these, it is most likely that the former was never really operational, reflecting the need to re-assert "legality" around the time of the 1974 Thai Binh Conference and the push to implement the NMS (Fforde 1984, 317).

15. In practice there clearly were conflicts of interest that varied in intensity. Procure-ment encountered resistance when the state had little to supply in return. The emphasis on harmony revealed the attitude of the group that laid down the formal structure presented in these documents.

16. That is, of troops and "socialist labor." I was told that "socialist duty-labor" was thirty days a year, and that the remuneration for this was deducted from the total worked in the individual's cooperative so as to, in effect, fine those who avoided it. In addition to the workpoints paid by the cooperative, any such work done for the state would bring 0.3 dong and some food (Hanoi I).

17. The meaning of Vietnamese terms marked with an asterisk is explained in the glossary.

18. Clauses 33 and 34. General prescriptions regarding village welfare may have had some effect, especially during the mid-1960s when decentralization during the U.S. bombings helped push educated cadres out of the towns. In the absence of reliable information, however, it is extremely hard to judge precisely the impact of these measures.

19. Vietnamese usage here differed radically from that in China, and the reader should avoid any confusion resulting from the use of similar English expressions. See "Note on Sino-Vietnamese terms" in the glossary.

20. "Weak" families—those with limited labor resources—would possibly be refused permission to join the cooperative by those who saw them as net claimants rather than positive contributors.

21. It is worth pointing out that the ultimate historical origins of the private plot in the basically Soviet model used here are to be found in Stalin's famous "pause" during early 1930. Declaring, in the midst of the "liquidation of the kulaks as a class," that there had been grave errors, he changed tack and made "important concessions to kolkhozy in relation to their animals and their household plots" (Davies 1980b, 261, 281). During the 1930s, of course, most Soviet members of kolkhozes—also higher-level agricultural pro-

ducer cooperatives—obtained essential supplies for subsistence from the private plots, while collective output went largely to the state as procurement and payments to the machine tractor stations. The classic study of the Soviet private plot is Wadekin (1973).

22. Note that this implied that only the cooperative could legally own land in the collective sector, and neither brigades nor households had more than usufruct to cooperative property.

23. As such they remained in principle individual or private property, as opposed to the "private" plots, which were collective (i.e., cooperative) property.

24. There were frequent instances in the histories of the cooperatives studied in detail below of cooperators wishing to leave in large numbers, but none was able to do so.

25. It should be emphasized that the family was distinct from the household (gia dinh). The former was basically a married couple with their children, if any. The latter was a unit of cohabitation. Thus a household could contain more than one family. This translation is roughly in accordance with contemporary English usage, but it is not universally followed, and, unfortunately, not in Fforde (1984). One problem with it is the fact that the standard polite enquiry, "Do you have a family?" in fact asks literally if you have your own household—"Co gia dinh rieng chua?" For obvious reasons, official population statistics are always given in terms of ho (or individuals). The ho was also the object of output distribution based upon workpoints earned plus rations for old people, and so forth.

26. Two additional points should be noted. First, there was no mention in any of the versions of the Statute of any specific interdiction of contracts with households that reportedly followed the so-called III-contracts controversy of the late 1960s (see Gordon 1972). Second, the subunits of the cooperative mentioned here did not have formal property rights to the assets allotted to them. Following Soviet practice, the economic system recognized three types of property-holding agent—state, collective, and individual—and the brigade was not one of these: it was not a "collective" unit (Hanoi I).

27. Hanoi I.

28. This paragraph is based on Hanoi I.

29. Clause 31. Note that there was no equivalent to the Chinese practice of allowing families an account in food-staples, which they could overdraw (Hanoi I; Parrish and Whyte 1978, 60).

30. These shares were slightly different in the 1974 Draft Translation—5–10 percent to the accumulation fund, 2–5 percent to the social fund.

31. In the system prescribed for the penalties and bonuses attached to the "III-point contracts" (hop dong ba chieu*) made with brigades, downside and upside risks differed: the "bonus for overfulfillment should never exceed 85 percent and only partly be paid in kind (in principle the cooperative has the right to fix this portion, but in practice 'it is a collective . . . .' [This circumlocution meant that, as a collective, the cooperative was subject to regulation by the state.] Penalties should be fixed beforehand, not surpass 50 percent, and be payable in cash" (Hanoi I). In practice there were considerable problems with the prescribed system because the high levels of free market prices made the penalties far lower than the rewards to deliberate underrecording of production by brigade leaders (Hanoi I).

32. Infantrymen: the "GI" of the North Vietnamese Army.

33. Hanoi I.

34. Ibid.

35. The discussion has already touched upon one of the most difficult analytical problems here, which was that economic functions had political implications, and vice versa.

36. In this discussion the "management team" of the cooperative is understood to include all members of the dual-level management structure—Management Committee, Supervisory Committee, brigade cadres, and others.

37. All means of production needed for collective production were collective property in the higher-level type of cooperative under discussion here.

38. The "socialist" nature (in the everyday sense of the word) of the cooperative depended upon the way in which this appropriated output was used, for in any production unit there must be some agency responsible for the distribution of output. In this case it was the brigade leadership that controlled collective output (at the harvest, if it was a crop).

39. The argument here is essentially static. Differences in the origins of authority also had their dynamic aspects. These are discussed further in chapter 3 in terms of the changing relative importance of control over output and control over production when technology is changing.

40. The great density of housing and the correspondingly low level of privacy made harmony in interfamily relations of considerable importance. Brigade cadres living in the same hamlet as the members of their brigade could therefore be subject to great pressure. Hanoi I emphasized the need for collusion if, for example, product was embezzled "because it is like one large family"; this would not preclude coercion of minor elements.

41. The importance of extended family obligations in creating abrupt demands for cash upon those capable of responding was obvious from direct observation in Hanoi during 1978-79.

42. My (currently untestable) impression is that wider kinship relations were probably the single most important unknown in attempting to understand social processes in any given cooperative.

43. A major problem with attempts to develop pig production was that a substantial proportion of the meat was eaten at meals given by cooperative cadres for cadres from regulatory agencies when they "went down" to the cooperative (Hanoi I). See chapter 8. This meant that far less was available for export. It also meant that selected social links were strengthened.

44. Comment from a female Vietnamese agricultural expert: "In the countryside, women must still endure (*chiu*) the men" (Hanoi I).

45. An implicit assumption here, which reflects North Vietnamese conditions at the time, is that although cash earned privately can readily be spent on consumer goods, the collective sector does not offer such opportunities. This was indeed generally true in North Vietnam, where collective distribution in cash was limited, primarily by the extremely low prices paid by the state for agricultural output. It is not a universal characteristic of collectivized agriculture.

46. The evidence from the incentive schemes in some of the cooperatives studied strongly suggested this.

47. See the problems encountered in trying to establish the situation in prereform My Tho cooperative (chapter 6).

Chapter 3

1. The problems involved in analyzing situations where property rights are only immanent have attracted interest from other disciplines, e.g., M. Gluckman (1955; 1972) and, on the Vietnamese legal system, G. Ginsburgs (1979). See also C. Humphrey (1983).

2. The idiomatic phrase used to describe situations where the brigades were the basic unit of collective activity was "work in brigades, eat in brigades" (*an chia doi**).

3. The idiomatic phrase used to describe this was "near land, far land; good land, bad land" (*co gan co xa, co tot co xau*)

4. It is important to distinguish clearly between autarkic self-reliance in respect to investment using physical inputs mainly produced within the cooperative (or brigade) and

that in respect to the generation of savings to finance investment. The first sense is used in the text both here and elsewhere.

Chapter 4

1. This name derives from the issue of *Vietnamese Studies*, edited by Nguyen Xuan Lai (no. 51, 1977), that discussed "The New Management System in a Cooperative."
2. Hanoi I.
3. These were to be "III-point contracts" covering the brigade's labor and nonlabor inputs and the output, augmented by some system of penalties and bonuses. There was no formal hierarchy of contracts in an essentially dual-level management system. Brigades did not make III-point contracts with other brigades for obvious reasons.
4. Hanoi I. Note that this method is in practice far less rational and scientific than it appears. In principle, the labor norms appear to provide an objective link between planned output, planned inputs, and actual output. But in practice the coefficients are neither stable nor predictable. There is no firm link between inputs and output. This is especially true in the production conditions examined here. North Vietnamese agriculture suffered enormous uncertainties because of the weather, pests, and other natural causes. In addition, there was no easily calculated relationship between the nominal and the effective labor input. Norm-based calculations resulted in a labor requirement expressed in man-hours (or a close equivalent). In the absence of coercion, nobody could determine how much would actually get done during that time, once the cooperator actually turned up.
5. This was a commonplace in contemporary pronouncements from the party: the simultaneous harmonization and satisfaction of "three interests" (*ba loi ich**)—individual, collective, and national.
6. For a detailed study of the policy toward the district, based upon fieldwork and experience in Vietnam, see Judith Appleton (1981). During the 1970s this policy was largely ineffective.
7. Le Duan (1975, 46).
8. Le Trong (1978, 54–56).
9. Le Trong, NCKT 100, pp. 53, 54.
10. Ibid., p. 54.
11. Ibid.; Nguyen Manh Huan (1979).
12. Hanoi I.
13. My informant spent some weeks in Vu Thang in 1979.
14. Le Trong (1978, 64).
15. Unless otherwise indicated, the following information on Dinh Cong comes from Le Trong (1978, 57–61).
16. This was partly because of extra traveling time since two brigades had now to be served (an extra twelve minutes on average per "half" of five hours in a ten-hour day), but mainly because of poor coordination, so that ploughmen would arrive before the rice had been harvested, the stubble not yet tidied or the water not yet correct. This cost another 12 percent of the time available. The conclusion was that "the organization of teams for ploughing with water buffalo during each ploughing season and in each brigade is the most profitable" (ibid., p. 57).
17. Thus, "this means that the brigade has joined the form of cooperation to a concrete division of labor" (p. 58).
18. In early 1970 Dinh Cong had experimented by allocating buffalo carts to three brigades: "to the three brigades of the three comrades—the secretary of the Party Committee, the president of the [People's Committee], and the manager of the cooperative" (p. 61). This was said to have led to a "sharp jump in labor productivity" and all the brigades wanted one.

19. The NMS sought to introduce a system of work-norms that were attached to the task, amplified by a categorization of labor that allowed the allocation of groups of workers to different sets of tasks, often with corresponding changes in remuneration (e.g., "strong" workers doing "heavy" work for higher pay). See *Études Vietnamiennes* 51:104–43.

20. Such calculations are made somewhat precarious by the practice of paying additional casual labor at harvest time by allotting points to their families via permanent laborers in the brigade.

21. The "mark" was a customary unit of cooperative accounting, usually equal to ten workpoints (see chapter 7).

22. This information came from the same source as note 11. Note that it contradicts the Statutory stipulation that the products of the household economy be "freely disposable."

23. Hanoi I.

24. Hanoi I.

25. For good historical reasons (e.g., Ho Chi Minh's early involvement with the Comintern), Vietnamese Communists were exposed to conventional Marxist-Leninist thinking from the Soviet Union and later China, with whose modes of expression they were probably (for reasons of cultural proximity) more comfortable. This enormously complicated area of research is far beyond the scope of the present study. Work such as D. G. Marr's *Vietnamese Tradition on Trial 1930–1945* (1981) is starting to open up the area of Vietnamese intellectual history. In my opinion work should also cover the wider social consciousness of the Vietnamese peasantry both before and after 1954.

26. This argument is presented in greater detail later—a schematic presentation is intended to highlight its basic structure.

27. It should be emphasized that cooperatives were seen as socialist economic organizations (see the definition of a cooperative in chapter 2).

28. Again (c.f. note 27) it should be emphasized that these specifically included agricultural producer cooperatives.

29. This recalls the Leninist slogan *"kto kogo"*—M. Lewin (1975, 359).

30. This understanding of the content of the technical revolution also gave an ideological basis for attacking "superindustrializers" in the discussions surrounding the First Five-Year Plan. If piecemeal advances in technology could be specifically said to be part of the revolution then this took some force from arguments that identified (in the Vietnamese context unrealistically) advance with heavy industry.

31. E.g., Le Duan (1969, 232).

32. Truong Chinh (1959, 24–29).

33. Le Duan (1969, 350).

34. Hanoi I. Emphasis added.

35. See Fforde (1986).

36. Unless otherwise indicated, this section is based upon Dinh Thu Cuc's excellent article in NCLS (1977) reviewing the cooperative movement.

37. Dinh Thu Cuc (1985, 30).

38. This detailed work provides fascinating insight into the early possibilities of cooperatives in North Vietnam before the U.S. war.

39. Fforde and Paine (1987, passim).

40. See Tran Duc Cuong (1979) for a discussion of the shift from lower- to higher-order cooperatives.

41. See the collection of articles, *Ve san xuat nong nghiep va hop tac hoa nong nghiep* (On Agricultural Production and Agricultural Cooperatives) (196).

42. In this year 95.1 percent of peasant family-units were in cooperatives, and 92.2 percent of peasant-cultivated land was collectivized (KTVH, table 69). According to the

same source, higher-order cooperatives had 94.3 percent of cooperator family-units, 83.6 percent of total cooperatives, and 91.6 percent of collectively cultivated land. The precise date of the figures quoted in the text from the 1969 report is unknown (quoted in Dinh Thu Cuc [1977, 40]). In 1969 there were 20,725 cooperatives, of which 3,402 were lower-order (KTVH, table 70).

43. KTVH, table 70. The source ostensibly refers to the total cultivated area but implicitly redefines it as the collectively cultivated area by using it to illustrate an argument about encroachment upon cooperative land.

44. The number of agricultural producer cooperatives fell from a maximum of 32,378 in 1964 to 17,000 in 1975 (KTVH, table 70).

45. The wider social implications of such a state of affairs in an area governed by a Marxist-Leninist party are clearly immense.

46. Hanoi II.

47. Nguyen Manh Huan (1980).

48. Nguyen Nien (1974, 10). Athough this article deals primarily with industrial management, the overall thrust of criticism was equally valid for agriculture.

49. Nguyen Tran Trong (1980, 29); Nguyen Lang and Nghiem Phu Ninh, in Nguyen Tri (1972, 122).

50. C.f. Decree no. 83-CP 2/5/75 of the Government Council CB 31/5/75, p. 126.

51. Circular letter no. 525-hd 23/6/75 of the State Arbitration Commission CB 15/7/75, p. 189.

52. Hoang Quoc Viet (1974, 39) quotes from the plenum resolution as follows: "We must struggle against the spontaneous nature of small-scale production and the customs and psychology of the old society."

53. Decree no. 75-CP of the Government Council 8/4/74 CB 15/4/74, p. 61.

54. A clear statement of pricing policy can be found in the Communique ("Thong Bao") of the Government Council (CB 30/9/74, pp. 226–27). The government clearly stated that it was not going to raise either basic or incentive prices paid for staples, although small marginal changes based upon regional differences were planned. A higher propertion of pork was to be bought at incentive prices. The incentive price for pork was subsequently set at 150 percent of the basic price—CB 31/10/74, p. 263.

55. Hoang Anh (1974). The author was then a secretary of the party's Central Committee and a vice-chairman of the Council of Ministers. The source used here is a French translation in photocopy circulated to diplomatic representatives. Criticisms similar to these were widespread.

56. Le Duan (1975, 44–45).

57. University of Hanoi "Course in Industrial Economics" (1979, 46).

58. I have not been able to find a copy of this decree. The basic party document defining the NMS line is Party Secretariat Decree 208-CT/TW of 16/9/74 (The Dat 1981, 211). This has also not been found.

Chapter 5

1. Photocopies of documents provided by Hanoi University in 1979.

2. See chapter 4 on the consequences for criticism of practice of the Vietnamese party's view of the Vietnamese revolution; without use of a means/ends dichotomy, criticism usually had to be expressed in terms of deviation from the party "ideal"—most commonly the NMS as realized in model cooperatives.

3. The former was a criticism made of Vu Thang before it became a model cooperative; the latter occurred in Van Tien (see chapter 9).

4. The average size of cooperative in the DRV in 1974–75 was around 190 families (KTVH, table 7).

5. This was based upon Dai Thang (Quynh Luu), Van Tien, Giao Tan, and Vu Thang around 1975. In the same year the average figure for the DRV was reported to be 16.5 percent (KTVH, tables 81, 93).

6. The relative absence of studies of nominal cooperatives in the published and otherwise available materials does not necessarily mean that inner-party research was not well informed as to the true state of affairs. But the available work was primarily intended to show cadres what they should have been doing, rather than function as "disinterested research." This meant, though, that authors—and therefore debate—had to focus on problems in implementing policy rather than broader issues.

7. See Houtart and Lemercier (1984) for a study revealing the extent of contemporary limitations on fieldwork.

Chapter 6

1. During 1973 My Tho's three smaller cooperatives were amalgamated into a larger one covering the entire commune. In 1974 a series of articles appeared in *Economic Research* (NCKT) entitled "Realizing a New Managerial Discipline in a Delta Cooperative," by Dang van Ngu (nos. 80, 81, 82, and 83). These gave a detailed critique of the prereform situation and described the progress of the reforms. It is not known why My Tho was singled out for publicity. It may be that the process of reform was unusual, in that it was carried out by the district rather than the province or center, and the picture painted was intended to justify this. Unless otherwise indicated, all quotations in this chapter are from this source, issue and page as cited.

2. Parts of the articles were translated and published for foreign consumption in *Vietnamese Studies* 51. Most of the deepest and most revealing criticisms were omitted.

3. The author of the articles seemed to be intimately involved with the affair. He gave an impression of concern with overcoming dishonesty and conservatism in order to put through something believed to benefit both society and cooperators. It was therefore possible to observe simultaneously not only just how far practice could diverge from the ideal prescriptions of the NMS, but also the attitudes and practices of reformers.

4. The aggregate figures were not simple to analyze, and it was necessary to cross-check and work backward and forward. Some data were relatively straightforward: the cultivable area at the end of 1973 was 413 hectares; at the beginning of April 1974 the population was 2,938. But the labor force was computed in three ways: (1) those within the age limits exempting them from obligatory labor—1,036; (2) those mobilized for work within the cooperative—984; (3) the "equivalent" labor force, including an allowance for the old and young—1,131. One labor equivalent (*lao dong da qui**) was defined as two old people or three young people (Hanoi I). There were fifty-two people in the cooperative working in the so-called free branches, which probably referred to members of families registered with the cooperative but working as independent artisans or petty traders.

5. Between 1968 and 1973 the reported average annual production of paddy was 945 tonnes with a maximum of 1,056 tonnes and a minimum of 800 tonnes. The associated annual productivity per worker reported ranged from 1,310 kg to 1,810 kg annually. These implied a labor force of 610 and 585 respectively, which was not surprising for the direct labor participation in agriculture in a commune of this size. The productivity of rice land was said to be around 3 tonnes per hectare, which implied a rice area of 325 hectares for the six-year period. This in turn implied that there must have been a substantial area of private gardens and plots over and above the 5 percent land, if the situation was legal (see chapter 2 on the statutory position). It is more likely that the ca. 100 hectares of cultivated land that was not put to rice was effectively controlled by the household sector. But the evidence here is weak, with no clear indication of the underlying trends.

6. Increases in the sown area were identified as a source of higher yields from the

earliest days of cooperativization. See, for example, Bui Huy Dap (1962).

7. The aggregate figures referred to the total level of paddy delivered to the state and therefore included all categories—tax, duty-sales, and sales at incentive prices. "The average proportion of production mobilized was 26 percent, reaching extremes of 28.9 percent and 19 percent" (80, p. 14).

8. This assumption was necessary because the best and worst years were not identified in any other way.

9. The average distribution figures possibly included an allocation for "priority families," which was usually meant to amount to 20–30 percent of collective consumption (Hanoi I). The data were for levels "guaranteed to cooperators" [sic] and should therefore arguably have included the "priority families." No mention was made of working-capital needs, nor of the accumulation funds. But these were negligible by comparison with the shortfall.

There was a faint possibility that the figures given for rice distribution were in milled rice rather than paddy (writer's notes unclear). This would have been contrary to the usual practice, and it is also unlikely because the usual conversion factor of ca. 70 percent gave a milled rice equivalent for the paddy availability for the period of 21.3 x 0.70 = 15 kg per month, which is less than 16 kg per month, the figure quoted for the level "guaranteed to cooperators."

10. These calculations were based upon reported production data which were probably a gross underestimate. Under the reformed accounting system reported output apparently jumped by over 50 percent in a year (see below). But note that, in a showpiece, overreporting of output was not unlikely.

11. See chapter 3 on the likely consequences of the nominal cooperative.

12. A literal translation of tu tuong*.

13. Although this procedure, if correctly enforced, might lead to the same outcome as a plan, the sense of the quote is that the plan was in effect used to prevent a cooperative-wide management of economic resources.

14. Here a nominal cooperative could be observed fulfilling extraeconomic functions.

15. This suggests that there were other groups in the shadows who gave them these "production papers" to sign. These may simply have been groups of cooperators who operated some sort of contract system that the source does not care to mention openly. Certainly we have here a hint of the way in which interests outside the brigades of a nominal cooperative carried out incursions into the brigade's formal field of activity. This points to the existence of alliances within the collectivity of great interest (e.g., who gave them these papers to sign?). The source unfortunately did not give more details.

16. The source could not be expected to be too explicit on this point because it was not just party members in the commune who were implicitly being criticized.

17. This again hinted at some underlying power structure (see note 15).

18. This implied that not all of the new Management Committee were outsiders.

19. This statement is surprising in the light of the apparent ease with which the new cooperative could borrow (see next section) and throws interesting light on the relative priorities of this particular Management Team.

20. These totaled 130,000 dong. When the cooperative was started it had a total capital of 383,000 dong, of which 270,000 dong had been lent by the state, 15,600 dong from "other sources," and only 97,400 dong from its own resources. No complete accounting statement is available. Fixed capital was valued at 317,000 dong. In the crucial winter of 1973–74 the cooperative had to borrow 292,880 dong from the State Bank, of which 75,300 dong was for liquid capital. In addition, 250,000 dong had to be borrowed from the cooperators (!) to buy manure (from the cooperators) and pay for labor; this was paid back after the harvest. It is not known on what the proceeds of the early borrowings were spent.

21. The relatively high price (estimated at 1.4 dong/kg) implied the existence of a

multiple-price system for "incentive sales," for the standard official incentive price was 0.9 dong/kg (Hanoi I).

22. Dang van Ngu, NCKT 82, p. 25. Comments in *Vietnamese Studies* 51 referred to a radical reorganization of soil policy.

23. There unfortunately was no information on details of changes in the methods of production that would help to clarify this astonishing increase in reported output. In a showpiece for reformers, overreporting would not be unexpected.

24. Hanoi I.

25. Dang Van Ngu, NCKT 82, p. 31.

26. The number of marks allocated to management was around 10 percent of the total (ibid., 83, p. 24).

27. This was partly related to a desire to avoid sales onto the free market (ibid., p. 23). It was also part of a far wider policy trend that was trying to determine levels of distribution in cooperatives. This was not in practice important in the cooperatives studied here, but surfaced as a major issue in policy debates just after the 6th plenum of 1979. See chapter 12.

28. The precise implications of this are not obvious. It is hard to see how lack of finance could have been a binding constraint to simple reproduction in a relatively self-contained system; it referred perhaps to conditions applied by regulatory agencies that were not revealed by the source. There is no mention in the materials of purchased input problems.

29. The plan also contained figures for transport and portage, irrigation and drainage, piecework rates for carpentry and masonry work, and so on (ibid., 82, p. 31).

30. There were three sorts of labor: Type A—280 days annually and 14 points a workday; type B—260 days annually and 13 points a workday; type C—220 days annually and 12 points a workday (ibid., 83, p. 20). The penalty-bonus system was "70:50"—70 percent of overfulfillment was awarded, 50 percent of underfulfillment was deducted (ibid., p. 24).

31. Imposing the job upon the brigade heads possibly meant that they had to seek advice from the better educated younger cadres in order to carry out the complicated calculations involved.

32. NCKT 81, p. 34.

33. The workday was often equal to a mark—10 workpoints (*cong diem**) if a relatively unsophisticated system was in use.

34. It was especially exasperating that the sources did not reveal the extent to which any increase in the average level of distribution compensated the richer brigades for any relative loss of income. In this cooperative the power of external agents may have been determinant, but in other areas (for whom the articles may have been intended as a guide), material compensation may have been more important. An additional area (also not clarified) was the circumstances surrounding the sharp increases in cadre incomes from the cooperative and any cash distribution to cooperators; here the large state loans could have played an important role.

35. Here it must be emphasized that the sources themselves largely ignored such factors in explaining the success of the "reforms," and so the analysis here inevitably remains limited.

36. For example, part of *Vietnamese Studies* 51 and the later article on the cooperative by Vu Trong Khai in NCKT 88, 1975.

Chapter 7

1. Pham Tran Thinh (1976, 21–22); discussion of Dai Thang, Xuan Thuy.

2. Ibid., p. 71. The prices at which these sales were made are unknown. 120 kg/sao is

equivalent to 3.33 tonnes per hectare (1 sao = 1/10 mau; 1 mau = 0.36 hectare); Van Tien cooperative was getting 7.1 tonnes per hectare of rice in 1974–75. It was accepted that yields would be higher on the private plots, and they reached 9 tonnes/hectare in Dai Thang (Xuan Thuy) compared with 6 tonnes/hectare for the collective (ibid., p. 12). Such "rents" cannot, therefore, be considered punitive. Unfortunately, the value of the workpoint in Van Tien was not known, so a direct comparison of returns to labor cannot be made.

3. Ibid., p. 2.
4. Hanoi I.
5. KTVH, tables 81, 93.
6. Hanoi I. Shortfalls in supplies from the state therefore influenced labor incentives.
7. The available statistics refer to "marketed produce" in terms that identify it as taking the form of the prescribed "commodity" exchange between the state and collective sectors of the economy.
8. Hanoi I.
9. Nguyen Xuan Thang (1976, 71). Emphasis added.
10. Hoang Anh Quoc (1976, 10).
11. This provided an interesting example (in a cooperative with weak management) of the distributional functions of the cooperative that allowed some of the staples delivered up by the brigades to be used for the support of communal incomes. The price at which these staples were sold was not known, but the use of discriminatory prices against nonparticipants in cooperative activity (no. 4) indicated the possibility of some impact upon production.
12. Ibid., p. 11. Prices were unknown.
13. Lac Ha Long, NCKT 93.
14. Hoang Anh Quoc (1976, 16).
15. Ibid., p. 15.
16. The labor-duty was 288 days for men and 276 days for women. This was equivalent to 316–362 labor-days for men and 276–331 labor-days for women. Because a man earned between 1.1 and 1.27 labor-days per day while a woman earned between 1.0 and 1.2 labor-days per day it was impossible for men to earn less than 10 percent above the minimum for women (so long as the value of the "labor-day" was constant for all cooperators). The precise differences between each category of laborer are unknown. (Vu thi Dau 1976, 7).
17. Ibid.
18. Ibid., p. 8. The precise meaning of "ration" (dinh suat*) was unspecified.
19. Nguyen Manh Huan (1979).
20. Nguyen Hoang (1979, 28); discussion of Thang Long cooperative.

Chapter 8

1. See chapter 2. Issues surrounding collective pig-rearing were a particularly clear example of the economic consequences of the price dualism characteristic of an aggravated shortage economy under North Vietnamese conditions.
2. These were usually state duty prices—Hanoi I.
3. Pham Bich Hanh (1976, 13).
4. Ibid., p. 14.
5. Hanoi I.
6. Pham Bich Hanh (1976, 9).
7. Tran Dinh Thien (1976, 8).
8. This could allow for the adoption of improved strains provided by the state. Details of availability in the sample are unknown.

9. Pham Bich Hanh (1976, 14–15).

10. Ibid., p. 17.

11. E.g., ibid.

12. Ibid., p. 18. Emphasis added.

13. Tran Dinh Thien (1976, 74).

14. According to Pham Bich Hanh (1976, 13), this was 2 percent of output, but according to Tran Quang Lam (1976, 15), it was 5 percent.

15. This was in Giao Hien, the commune or village where the pigs had been kept before they were moved—Pham Bich Hanh (1976). The following discussion of Dai Thang is based on this source, p. 11.

16. Hanoi I mentioned this as a more general problem, revealing the low priority given in practice to collective pig-rearing.

17. There were a number of unfinished sties in this cooperative because they had originally planned to rear 1,000 pigs. Pham Bich Hanh (1976, 22).

18. Ibid., p. 8.

19. Ibid., p. 9. Note that the cooperative's management was aware of this but did not stop it.

20. Nguyen Manh Huan (1976). The duty was dependent upon the number of workers in the family; one worker, 3.6 tonnes; two, 6.4; three, 9.

21. Vu thi Dau (1976, 8). The land was "mainly used for growing rice."

22. *Sic*—this was presumably some "equivalent measure."

23. Tran Dinh Thien (1976, 71).

Chapter 9

1. The term literally means "trade branch" or "handicraft branch," but it has acquired the technical meaning in discussion of cooperative management problems translated here as "subsidiary branch."

2. Men tended to wear the factory-made sun hat—personal observation.

3. Cao van Chinh (1976). The discussion of Van Tien is based on this sources, pages as indicated in the text.

4. "It was a place famous for carpentry and masonry"—ibid., p. 2. Bricks and tiles had also been produced.

5. Literally, "to work outside" (*lam ngoai**). See the discussion of the "outside economy" in chapter 1.

6. Nguyen Xuan Thang (1976, 75).

7. Ibid., p. 72.

8. Vu thi Dau (1976, 11).

Chapter 10

1. See the discussion of household contracting in chapter 12.

2. Note that while no. 1 was concerned with the effectiveness of the control over labor, nos. 2–5 implied changes in labor allocation accompanying more centralied managerial control.

3. Lac Ha Long (1976).

4. In Vietnamese, "*gat den dau, cay do.*" See Glossary.

5. "*Bo doi trong xu*" (*sic*).

6. Vu Duc Thanh (1976, 15).

7. Labor mobilization during staples harvesting was easier than for other crops. In Dai Thang (Xuan Thuy) the jute was "taking too much labor" at harvest time; the crop was very sensitive to timing but coincided with the rice harvest—a delay of five to ten days

could reduce yields by 50 percent. But the cooperative was seen as having large quantities of "surplus labor." Pham Tran Thinh (1976, 8).

8. Lac Ha Long (1976).

9. "Ploughing and transplanting together are the determining and most basic link in the process of production." Pham Tran Thinh (1976, 18).

10. See the articles by Le Nhat Quang (1976) and Vu Tuan Anh (1974) for examples of technological determinism in full cry, where the size of production units is related entirely—and unjustifiably—to indivisibilities in various means of production.

11. *Tang vu*—increasing the number of harvests.

12. This illustrated the practical distinction between formal ownership and possession.

13. Casual observation of the delta countryside showed that the animals were tended by small boys in the traditional manner.

14. Tran Dinh Thien (1976, 70). Emphasis added.

15. Detail: percentage of work done with draft animals in Van Tien in 1974–75: Ploughing, brigade 1, 65 percent; brigade 2, 45 percent. Harrowing, brigade 1, 75 percent; brigade 2, 45 percent. Vu Duc Thanh (1976, 17).

16. 1 sao = 0.036 hectare.

17. Nguyen Bich Huong (1976, 10) referred to this brigade as a team. Since it was directly run by the Management Committee the difference is unimportant and a problem only of terminology.

18. This was again expressed by the expression *gat den dau, cay do** for which there is no simple English equivalent.

19. Forty-one days were available for preparing for the Vth month crop and thirty-eight for the Xth. Lac Ha Long (1976).

20. Again, *tang vu*—increasing the number of harvests.

21. No mention was made of any possible need for additional "outside" payments to tractor drivers to ensure proper access to their services.

22. Lac Ha Long (1975).

23. Tran Quang Lam (1976).

24. The sources did not differentiate between improved local varieties and new strains.

25. Nguyen Bich Huong (1976, 7–8).

26. Ibid., pp. 8–9.

27. Tran Quang Lam (1976, 12).

28. Nguyen Hoang (1979, 28).

29. The fertilizer brigade was merely an organizational form that grouped a number of teams that were each part of a rice brigade. It had little more than a nominal existence. Tran Dinh Thyien (1976, 70).

30. Ibid., p. 71.

31. Pham Tran Thinh (1976, 14).

32. Tran Quang Lam (1976, 13).

33. According to Hanoi I this usually absorbed the larger part of the manure generated by the collective herd.

34. I.e., of the order of 600 tonnes annually; with 1,000 pigs in the collective herd this was most of the likely supply.

35. Lac Ha Long (1976).

36. Ibid., p. 51; Tran Quang Lam (1976, 12).

37. Pham Tran Thinh (1976, 14).

38. Ibid.

39. Tran Quang Lam (1976, 13); Nguyen Bich Huong (1976, 10); Le Trong (1978, 58).

40. *"Doi thuy loi," "doi thuy nong."* To my knowledge these are synonyms.

41. The other was a *doi nong giang* organized into small groups or individuals responsible for fields.

42. By limiting land-scattering this presumably reduced the number of brigades having particular interests in any single part of the water-control system.

43. Two, not one, could be used because of the amalgamated cooperative.

44. Tran Quang Lam (1976, 6).

45. Ibid.

46. Tran Dinh Thien (1976, 73).

47. This stemmed partly from the preponderance of private property in land and the correspondingly small size of production unit in the particular historical context. The important economic and social functions of the commune were not able to offset this.

48. Pham Tran Thinh (1976, 15).

49. The effects of irrigation works upon collective incomes may have been more clearly perceived because the system of "work in brigades, eat in brigades" had been abolished.

50. This was one of the clearest examples available of the problems involved in managing units when there was no clear correspondence between production conditions and perceptions of group interests.

51. Tran Quang Lam (1976, 6).

52. Nguyen Hoang (1979, 26).

53. Nguyen Manh Huan (1979, 43).

54. Le Trong (1979, 63). Emphasis in original.

55. Nguyen Bich Huong (1976, 7).

Chapter 11

1. Francesca Bray (1983) provides a valuable comparison of the differing scale economies of rice and other crops, which has important implications for the possible economic value of collectivization.

2. See Fforde (1983) for a more detailed exposition of the historical background to agricultural collectivization in North Vietnam.

3. See the discussion of "observation theories" in Lakatos (1970).

4. For a powerful argument about the inherently unpredictable nature of human "rule-governed" behavior see Winch (1963); for a characteristically robust North Vietnamese view of the "ocean of knowledge" see the quote from Hoang Kim Giao at the end of the preface to this volume.

5. It is possible to argue that one of the great dilemmas confronting socialist policymakers is that, in the absence of widespread commoditization, calculation of the social value of economic activity is extremely difficult.

Chapter 12

1. Fforde (1984a) attempts to model this process formally.

2. Free-market rice prices in 1974 were around 0.8–0.9 dong/kg paddy but had risen to around 4 dong by 1979. Ngo Vinh Long (1985) and personal observation.

3. There are no easily accessible data on procurement in the North alone. National data on output and procurement can be found in table 12.2, showing that total domestic supplies to the state fell over 30 percent between 1976 and 1979.

4. Ngo Vinh Long (1984); Lam Thanh Liem (1984).

5. Fforde (1985) looks at Soviet-Vietnamese economic relations and evidence for Soviet influence on the policy changes from 1979 to 1982.

6. See Circular Letters 291-TTg of the Prime Minister's Office in CB 15–5–78 and 3-NN/CV/CN of the Ministry of Agriculture in CB 15–7–78.

7. Le Trong (1980b, 26) refers to average levels of monthly collective distribution in agricultural cooperatives of 5–6 kg of paddy. The concern to maintain procurement levels comes out clearly in the regular government decrees relating to procurement, such as Prime Minister's Orders 154-TTg "On Staples Work in the 1974 Winter-Spring Harvest" (17–6–74 in CB 29–6–74); 275-TTg "On Staples Work in the 1974 Xth Month Harvest and Implementing Widely and Thoroughly Decree 75-CP of 8–4–74" (1–11–74 in CB 15–11–74); Order 85-CP of the Government Council HDCP "On Staples Work in 1975" (6–5–75 in CB 31–5–75); Prime Minister's Orders 240-TTg "On Staples Work for the 1976 Vth and Spring Harvests" (1–6–76 in CB 15–6–76); and 287-TTg "On the Stimulation of Emulation so as to Raise Staples Output . . . for a Successful Xth Month Harvest" (18–7–77 in CB 31–7–77).

8. See CB 8–1–78.

9. The Dat (1981, 129).

10. NhD 20 and 21–6–79.

11. NhD 21–6–79. Emphasis added.

12. NhD 10–3–82, in the runup to the Fifth Congress, explicitly stated that the Sixth Plenum had only lasted for one day—August 13—and further omitted any reference to discussion of agrarian policy.

13. Ibid., p. 18: "Manh dan sua doi va cai tien cac chinh sach hien hanh, nhat la ve luu thong, phan phoi, nham lam cho san xuat bung ra dung huong" (Strongly Correct and Reform Current Policies, Especially Those Regarding Circulation and Distribution, in Order to Make Production Burst Out in the Right Direction).

14. CB 15–9–79. The decree stressed the need for provinces and cities to adopt ways of ensuring that the area sown in the forthcoming winter crop was maximized. Output on land "lent" (muon) to cooperators in this way was not to become part of collective distribution (khong tinh vao phuong an an chia). Cooperatives were to help the direct cultivators, especially in the links of seeds, ploughing and harrowing, and irrigation. Similar regulations were to apply to land that had moved out of cultivation, although here land that had long been uncultivated was to be "lent" for up to five years depending upon the work needed to bring it back into cultivation. Such land was no longer to be subject to any tax.

15. Government Council Decree 357-CP, "Policies to Stimulate the Development of Livestock-Rearing" (3–10–79 in CB 15–10–79). This can also be found in the BBC's "Summary of World Broadcasts," 31–10–79.

16. A translation of the decree can be found in the BBC's "Summary of World Broadcasts," 21–11–79.

17. Ngo Vinh Long (1985, 46).

18. Decree 373-CP 13–10–79 in CB 31–10–79.

19. This resolution (26-NQ/TW of 23–6–80) stressed the need to expand socialist trade in staples and encourage private trade to "move in the right direction." It heralded a number of operations against the free market that appear to have been relatively successful. For details of this resolution see NhD 24–3–82.

20. It is interesting to note the parallels between the ways in which the NMS had sought to isolate the brigades (see the quote from Lac Ha Long in chapter 10) and the attempt to reserve certain links for the cooperative under the output contracting system.

21. Nguyen Khanh Toan (1980, 32–33). This was his speech at a "Scientific Conference to Research the Sixth Plenum."

22. Le Trong (1980a, 61).

23. Le Trong (1980b, 27–30).

24. Le Trong (1980c, 35).

25. In January 1980 the state Social Sciences Commission had organized a "Scientific Conference to Research the Sixth Plenum," and papers from that meeting were subsequently published (State Social Science Commission 1980).

26. Nguyen Huy (1980, 12).

27. Huu Hanh (1980, 41).

28. *Doan ket* (Paris) 18-4-1981, quoting *Dai doan ket*.

29. Apparently Secretariat Communique no. 22-TB/TW of 21-10-80 "summarized the results of current work in the reinforcement of agricultural cooperatives hand-in-hand with the construction of districts in Northern and Central delta areas, of which contract work, especially contracts with cooperators, was especially important." Le Thanh Nghi (1981, 33). Order 100-CT/TU was entitled "Reform of contract work and the expansion of output contracts with labor groups and laborers in agricultural cooperatives."

30. See Le Thanh Nghi (1981, 32-79). This pamphlet must be considered one of the most important sources on the output contract system. The following discussion is based on this source, pages as indicated in the text.

31. Order of Government Council 372-CP 10-10-79, CB 31-10-79 "On Urgent Measures to Stimulate Production and Savings of Staples in the Entire Country."

32. Government Council Decree 09-CP of 9-1-80 "On Staples Policy" announced the intention to stabilize procurement levels (CB 15-2-80, p. 61). This was then followed by Decree 310-CP of 1-10-80 "On the Policy of Stabilizing the Duties of Agricultural Cooperatives, Production Collectives, and Peasant Families to Sell Agricultural Produce to the State" (CB 15-10-80). This stipulated clearly that "apart from the duty to contribute and sell [to the state], the producer has the right to use and dispose of remaining output. . . . State trade will purchase [it] at negotiated prices if necessary" (CB 15-10-80, p. 294). This, however, contradicts the spirit of the simultaneously issued Decree 312-CP, "On the Reinforcement of Market Management" (CB 31-10-80), which restated the state's monopoly rights to organize trade in various goods, including staples (CB 31-10-80, p. 322).

33. See note 32.

34. Ngo Vinh Long (1985, 54 et seq.) suggests that the brigades administered the contracts; this contradicts Nguyen Huy and Huu Hanh above.

35. This view, with which the present writer provisionally agrees, is also that of Ngo Vinh Long (1985). The wide differences between individual cooperatives revealed by the discussion in the earlier part of the book provides considerable prima facie support for the idea that there would be a great variation in response to the new official policy.

36. This view is typified by Werner (1984).

37. Staples output data for individual provinces are not available for all years, 1976–1982. Crucially, 1979 is lacking. Total staples output in million tonnes of paddy equivalent for the eleven key Northern delta or semidelta provinces/cities (Vinh Phu, Ha Bac, Quang Ninh, Hai Hung, Thai Binh, Ha Son Binh, Ha Nam Ninh, Thanh Hoa, Nghe Tinh, Haiphong, and Hanoi) reportedly moved as follows: 1976, 5.2; 1980, 4.6; 1981, 5.6; 1982, 5.9; 1983, 5.8; 1984, 5.7. SLTK (1981, 1982, and 1984). It should be noted, however, that most of the output gain in 1979–80 probably took place in the Mekong Delta, where total staples output rose by 0.60 m. tonnes (from 4.90 to 5.50), compared with a national increase of 0.65 m. tonnes (from 14.38 to 15.07). Data from *Tap chi thong ke* (Statistics) 3 (1983), table 4; SLTK (1982, table 32).

Glossary of Vietnamese Technical and Vernacular Terms

Note on Sino-Vietnamese Terms

The table below is intended to provide clarification of the differences and similarities between Vietnamese and Chinese Communist practice in the use of technical terms. For a fuller exposition the reader should consult the relevant dictionaries.

Vietnamese/Mandarin Chinese
(Conventional English translation)

xã/she
(Commune) (Organized body, commune, god of the land)

hợp tác xã/hezuoshe
(Cooperative) (Cooperative)

xã viên/sheyuan
(Cooperator) (Commune member)

công xã/gongshe
(Commune) (Commune)

đội/dui
(Brigade) (Team)

đại đội/dadui
(Great brigade) (Brigade)

hộ/hu
(Family) (Door; household, family)

gia đình/jiating
(Household) (Household, family)

Source: Bùi Phụng, *Từ-điển-Việt-Anh* (Hànội, 1978); *The Pinyin-Chinese-English Dictionary* (London, 1979).

Note: It is of interest that the usual Chinese word for administrative village (*xiang:* in Sino-Vietnamese, *hương*) has apparently passed out of Vietnamese administrative usage.

Glossary

ăn chia đội   Idiomatic phrase used to describe brigade-level distribution. A technical equivalent (e.g., *phân phối theo đội*) was probably avoided in order to emphasize the wider behavioral implications for brigade independence—strong brigade usufruct to land and other assets. Here translated (inadequately) as "work in brigades, eat in brigades," attempting to retain some of the idiomatic flavor.

ba lòi ích   Three interests—refers to the need to harmonize the interests of the state, the collective, and the individual in any given activity.

bậc   The grades into which working members were divided in some cooperatives to determine the value of the work done—the points allocated the cooperator.

ban kiểm tra   The Supervisory Committee of the cooperative.

bộ đội   Soldier (literally, infantry-squaddy)—ordinary soldier in the People's Army.

cải cách ruộng đất   Land reform—with social sense, i.e., land reallocation to different holders.

cải tạo đất   Literally land reform, referring to the activities such as land leveling and canal digging that helped prepare the land for new seeds and cropping patterns: land improvement.

chăm sóc   Caring for plants (a literal translation).

chi bộ   The Party Committee which corresponded to the village or commune and grouped together the party members of the administrative unit.

chủ nhiệm ban quản trị   The chairman of the Management Committee of the cooperative, its manager.

chủ tịch xã   The commune president, the head of the People's Committee in the commune.

có gần có xa, có tốt có xấu   "Near land—far land, good land—bad land," refers to one of the consequences of brigade autonomy: land scattering, where each brigade's land was scattered about the commune. It is thought to reflect the pattern of private land allocation after land reform but before cooperativization.

công điểm   According to Bùi Phụng, this was "cooperative mark for work done." Translatable as workpoints for a standard day, usually equal to ten points, or workpoints per day.

công trình thủy lợi   Water works—refers to all investment projects designed to improve the control of water in agriculture, for both irrigation and drainage.

đại hội xã viên   The General Assembly of a cooperative, in principle its highest organ.

đất năm phần trăm   The 5 percent land, which was directly worked by cooperators. The term was also often used to mean the land of the cooperative that was farmed by the cooperators of their own account (the "private plots," excluding privately owned land) and could therefore exceed the legal limit of 5 percent of the per capita holdings of the cooperative per family unit.

di chạy hàng xén   An idiomatic phrase used to describe unwanted enthusiasm on the part of the cooperators for the pursuit of petty-commodity production and trading activities, translated as "run to the market."

điều lệ hợp tác xã   Cooperative Statutes—the draft Statutes issued, with prescriptive force, to govern the structure and workings of cooperatives.

đình   The communal house of the traditional commune, center of religious and social ranking rituals.

đinh suất   Ration. Unit of distribution in some cooperatives.

đội   The brigade, the largest subunit of a cooperative.

đội phó   The deputy leader of a brigade.

đội trưởng   Brigade head.

đồng   The basic unit of Vietnamese currency. Monthly wages for an urban worker were of the order of 40–60 đồng during the period studied here. Divided into ten *hào* and 100 *xu*.

gặt đến đâu, cầy đó   An idiomatic phrase used to describe an efficient coordination of ploughing and harvesting so that neither fields nor plough teams were left idle. Translatable literally as "wherever the harvesting has finished, plough there," where the flavor of the original is absent.

gia đình   Household, coresident unit: the basic social unit. Note the basic polite question "Anh chị có gia đình riêng chưa?" (Do you have a family of your own yet?)—to which a feasible answer could be "I am married" as well as "I have children."

giá khuyến khích   Incentive price: the price at which so-called incentive sales to the state were made. Precise details of procedures unknown. Incentive prices differed according to circumstances.

giá nhiệm vụ   Duty price: the price at which so-called duty sales to the state were made. Precise details of procedures unknown.

họ   Family, nuclear family: the basic administrative unit. Population data usually given in terms of these units.

hội đồng kỳ mục   The Council of Notables of the traditional commune.

hợp đồng ba chiều   Three-point contract: the contract between the Management Committee and a subunit of the cooperative (usually a brigade) that stipulated the level of (1) labor inputs, (2) nonlabor inputs, and (3) output, or the quantity to be delivered to the Management Committee.

hợp đồng ba chiều với thưởng phạt   Three-point contract with penalties and bonuses applied to the brigade by the Management Committee for nonfulfillment of the contract.

hợp tác xã   Cooperative—literally, united-task-commune.

hợp tác xã sản xuất nông nghiệp bậc cao   Literally, higher-level agricultural producer cooperative. In lower-level cooperatives some collective income was distributed according to the assets contributed by the cooperator; this did not happen in the so-called higher-level cooperative.

huyện   District—the administrative district between the commune and the province; as such it was the level immediately superior to the important commune-level organizations of the party and the commune-level cooperatives.

khâu   Literally, link, a word with a similar meaning to the English (i.e., the links in a chain). The constituent tasks into which a labor process could be analyzed.

kinh tế phụ gia đình   The minor or subsidiary family economy—own-account activities by cooperators and their families.

làm đất   Land preparation (literally—making land): ploughing, harrowing, and preparing for transplanting.

làm ngoại   To work outside the state or collective sectors. The word *ngoại* is semnatically broad (antonym—*nội*).

lao động dã qui   Equivalent labor. A measure of the labor supply of a cooperative that included allowances for minor labor.

lao động rồi rào   Superfluous or superabundant labor.

lúa chiêm   Vth month rice (literally—Cham rice), the traditional minor rice crop transplanted in January and harvested in May/June.

lúa mùa   Xth month rice (literally—the harvest), the traditional major rice crop transplanted in July/August and harvested in December.

lúa thu   Autumn rice, a literal translation.

lúa xuân   Spring rice, a literal translation. A more rapidly maturing rice that replaced Vth month rice as part of a move to increase harvests.

lý trưởng   The mayor of the traditional commune. According to Phan Khoang (1966), *lý* here is an old synonym for *làng* (often equivalent to *thôn*).

mẫu   A unit of measurement usually equalling 0.36 ha. in the North, 0.50 ha. in the Center.

ngành   Branch—production within cooperatives was divided into three areas, known as branches: cultivation—*trồng trọt*; livestock—*chẵn nuôi*; subsidiary branches—*ngành nghề*.

ngày công   Work-day or labor-day; the term used to describe the activity required of a cooperator to fulfill a day's norm.

nhiệm vụ   Duty and/or task. The word unites the different senses of duty and duties in referring to both moral and the contractual aspects.

nhóm   Group—an ad hoc and small unit of organization of labor in production.

quỹ   Funds—deductions from the collective income of a cooperative after deliveries to the state have been made and before the distribution of incomes to cooperators. Two major sources of these deductions were:

quỹ công ích   The social fund; and

quỹ tích lũy   The accumulation fund.

quyết định   Literally, to determine the outcome of some process or sequence of events. In practice it often merely meant "to have a strong effect upon." Translated here (euphemistically) as "determine."

tăng vụ   Increasing harvests (an inadequate translation)—this referred to the attempt to increase land yields and end rice monoculture by introducing a winter crop of nonrice staples or an industrial crop while shortening the growing period for the minor rice crop by replacing Vth month rice with Spring rice. It could also refer simply to the introduction of a second rice crop in an area where it had not previously been grown.

thi đua   Emulation—literally, competitions to increase labor productivity.

thôn   Village—the constituent unit of the commune.

thuế nhà nước   State taxes—the priority procurement category, in principle fixed by reference to the land yields of the cooperative. In bad years it could be reduced (*miễn*).

tiên tiến   Advanced—literal translation used to refer to cooperatives that, although not sufficiently close to Party prescription to be called model and be made a subject for emulation campaigns, were successful enough to warrant study.

tỉnh   Province—the level superior to the district and effectively just below the center—Hanoi.

tổ   Team—a regular unit of organization of labor in production, either as a constituent unit of a brigade or as a unit managed directly by the Management Committee.

trung du   The midlands—land outside the deltas but below about 500–1,000 m. Cooperatives on the borders of the deltas often had land in this region.

trung ương   The center—an idiomatic phrase referring to the national level of state and party, e.g., the Central Committee of the party, or the relevant ministry. Often contrasted with *địa phương*—regional.

tư tưởng   Ideology (literal translation)—also has the meaning of "confidence in the line of the party" as in "to lose ideology" (*mất tư tưởng*).

xã   Commune—the administrative-geographic level that traditionally formed an intermediary between state and peasantry.

xã viên   Cooperator (literally, member of a commune)—derives from the Vietnamese for cooperative—*hợp tác xã*.

xóm   Hamlet—the smallest level of geographic groupings of population.

Bibliography

Notes on Source Materials

Interviews were obtained during the academic year 1978–79 with staff of the following:
—Economics and Politics Faculty (*Khoa Kinh Te Chinh Tri*), Hanoi University
—History Faculty (*Khoa Lich Su*), Hanoi University
—Economic Research Commission (*Vien Kinh Te Hoc*)
—Historical Research Commission (*Vien Su Hoc*).
In addition, a number of visits were made to "showpiece" cooperatives near Hanoi. All interviews were carried out in Vietnamese without the assistance of an interpreter. Notes collected from these discussions were grouped into two areas covering nonnumeric and numeric data; these are referred to respectively as "Hanoi I" and "Hanoi II." In addition, I was allowed access to a series of studies written by students at the Economic and Politics Faculty that were the product of their visits to agricultural cooperatives as part of their practical training. These covered various aspects of developmental problems in the cooperatives and are marked as University of Hanoi material (UH) below. Photocopies are in my possession.

The main source of aggregate statistical material, apart from the collections referred to under "Hanoi II," is *Kinh te va Van hoa Viet Nam* (Economy and Culture of Vietnam) (Hanoi: Tong cuc Thong Ke, 1978. This is referred to throughout as KTVH.

Table B1 groups the primary sources used to build up information on the sample of cooperatives.

Table B.1

Individual Cooperatives

Dai Thang, Xuan Thuy	Pham Bich Hanh (UH), "Ve van de phat trien chan nuoi o hop tac xa Dai Thang" (On the Problem of Developing Livestock in Dai Thang).
	Nguyen Bich Huong (UH), "Mot so bien phap ve van de day manh nganh trong trot o hop tac xa Dai Thang" (Some Principles Concerning the Problem of Developing Production in Dai Thang Cooperative).
	Pham Tran Thinh (UH), "Su dung hop ly nguon lao dong o hop tac xa Dai Thang" (On the Rational Use of the Labor Supply in Dai Thang Cooperative).
Dai Thang, Quynh Luu	Vu thi Dau (UH), "Van de su dung lao dong nong nghiep o hop tac xa Dai Thang" (The Problem of Agricultural Labor Use in Dai Thang Cooperative).
Dinh Cong	Le Trong (NCKT 101, 1978), "Kinh nghiem ve su phat trien cac hinh thuc to chuc lao dong cua hop tac xa Dinh Cong" (Experiences with the Development of Forms of Production in Dinh Cong Cooperative).
Giao Tan	An Quoc (UH), "Ve van de phan phoi ve thu nhap cua hop tac xa Giao Tan" (On the Problem of Distribution and Income in Giao Tan Cooperative).
	Nguyen Xuan Thang (NCKT 94, 1976), "Nghe phu o hop tac xa Giao Tan" (Subsidiary Branches in Giao Tan Cooperative).
	Tran Quang Lam (UH), "Ve van de mo rong qui mo cua hop tax xa Giao Tan" (On the Problem of Expansion in Giao Tan Cooperative).
Hai Nam	Lac Ha Long (NCKT 93, 1976), "May van de ve quan ly va su dung tu lieu san xuat o hop tac xa Hai Nam" (Some Problems in the Organization and Use of Means of Production in Hai Nam Cooperative).
My Tho	Dang van Ngu (NCKT 80, 81, 82, and 83, 1974), "Thuc hien nen nep quan ly moi o mot hop tac xa vung dong bang" (Realizing a New Management System in a Delta Cooperative).
	Vu Trong Khai (NCKT 88, 1975), "To chuc san xuat va cai tien quan ly o hop tac xa nong nghiep My Tho" (The Organization of Production and Management Reform in Agricultural Cooperative My Tho).
Thang Long	Nguyen Hoang (NCKT 108, 1979), "Hop tac xa nong nghiep Thang long dua nong nghiep len san xuat lon xa hoi chu nghia" (Thang Long Cooperative Leading Agriculture to Large-Scale Socialist Production).
Van Tien	Cao van Chinh (UH), "Vai van de ve phat trien nganh nghe o hop tac xa Van Tien" (Some Problems in the Development of Subsidiary Branches in Van Tien Cooperative).
	Tran Dinh Thien (UH and NCKT 91, 1976), "May van de ve phan cong lao dong o hop tac xa Van Tien" (Some Problems of Labor Allocation in Van Tien Cooperative).

Table B.1 (continued)

Vu Duc Thanh, "May y kien ve nganh trong trot o Van Tien" (Some Opinions on the Cultivation Branch in Van Tien).

Vu Thang   Nguyen Manh Huan (NCKT 107, 1979), "Hop tac xa Vu Thang" (Vu Thang Cooperative).

Nguyen Manh Huan, "Vu Thang Agriculture Cooperative, an Advanced Typical Cooperative of Vietnam," UNITAR International Conference on Alternative Development Strategies and the Future of Asia, New Delhi, March 1980.

The following abbreviations are used in the bibliography:

BEI   *Bulletin économique indochinoise*
BEFEO   *Bulletin de l'École francaise de l'extrême Orient*
BSEI   *Bulletin de la Société des Études indochinoises*
NTVNTLS   *Nong thon Viet nam trong lich su* (The Vietnamese Countryside in History). Hanoi, 1977.
RI   *Revue indochinoise*

Annuaire Statistique de l'Indochine, various years. Hanoi.
Appleton, J. 1985. "District Debate in Vietnam 1970–80: Towards Decentralisation," B.A. thesis, University of East Anglia, 1981.
Baffeluf, A. 1910. *Les impots en Annam*. Paris.
Baker, H. D. R. 1979. *Chinese Family and Kinship*. London.
Ban nghien cuu su, dia, van (Research Committee for History, Geography, and Culture. "Cach mang thang tam va van de ruong dat" (The August Revolution and the Land Question). In special issue of *Tap san su, dia van* (History-Geography-Culture: Studies) 2.
Benedict, P. 1947. "An Analysis of Annamite Kinship Terms," *Southwestern Journal of Anthropology* 3.
Bernadini, J. J. 1931. *Sous la botte nippone*. Paris.
Bernard, P. 1934. *Le problème économique indochinois*. Paris.
———. 1937. *Nouveaux aspects du problème économique Indochinois*. Paris.
Betts, R. F. 1961. *Assimilation and Association in French Colonial theory 1890–1914*. New York.
Bhaduri, D. 1973. "A Study of Agricultural Backwardness under Semi-feudalism," *Economic Journal* (March).
Boudarel, G. 1968. "Memoires de Phan boi Chau," *France-Asie* 4.
———. 1969. "Phan boi Chau et la societe vietnamienne de son temps," *France-Asie* 1.
Boudillon, A. 1915. *Le régime de la propriété foncière en Indochine*. Paris.
Bouinais, A., and A. Paulus. 1885. *Le protectorat du Tonkin*. Paris.
Bournier, P. 1925. "Études sur la consommation du riz en Indochine," BEI.
Bray, F. 1983. "Patterns of Evolution in Rice-Growing Societies," *Journal of Peasant Studies* (October).
Brunschweg, H. 1966. *French Colonialism, Myths and Realities 1871–1914*. London.
Bui Huy Dap. 1962. *Cay lua mien bac Viet Nam* (Rice Cultivation in North Vietnam). Hanoi.
Bui Phung. 1978. *Tu dien Viet-Anh* (Vietnamese-English Dictionary). Hanoi.
Buttinger, J. 1958. *The Smaller Dragon*. New York.

————. 1967. *Vietnam: A Dragon Embattled*. New York.

————. 1969. *Vietnam: A Political History*. London.

Cadiere, L. 1958. *Croyance et pratiques religieuses des Vietnamiens*. Saigon.

Cady, J. F. 1964. *Southeast Asia, Its Historical Development*. New York.

————. 1967. *The Roots of French Imperialism in East Asia*. 2d ed. Ithaca.

Cameron, R., ed. 1974. *Essays in French Economic History*. London.

Cao van Chinh. 1976. "Vai van de ve phat trien nganh nghe hop tac xa Van Tien" (Some Problems in the Development of Subsidiary Branches in Van Tien Cooperative). UH.

Caron, F. 1974. "French Railroad Investment 1850–1914." In *Essays in French Economic History*, ed. R. Cameron. London.

Chaigneau, M. D. 1867. *Souvenirs de Hue*. Paris.

Chassigneux, E. 1912. "L'irrigation dans le delta du Tonkin," *La Géographie*.

————. 1927. "La plaine et les irrigations de Thanh-Hoa," *Annales de Géographie*.

Chen, N. C. 1966. *Chinese Economic Statistics*. Edinburgh.

Chesneaux, J. 1961. "French Historiography and the Evolution of Colonial Vietnam." In *"Historians of Southeast Asia,"* ed. D. G. E. Hall. London.

————. 1966. *"The Vietnamese Nation*. Sydney.

Coedes, G. 1966. *The Making of Southeast Asia*. Berkeley.

Congrès internationale de la population. 1937. Paris.

Cotter, M. G. 1968. "Towards a Social History of the Vietnamese Southwards Movement," *Journal of Southeast Asian History* 1.

Crawfurd, J. 1830. *Journal of an Embassy from the Governor-General of India to the Courts of Siam and Cochinchina*. 2d ed. London.

Dang lao dong Viet Nam (Vietnam Workers' Party). 1959. "Nghi quyet cua hoi nghi trung uong lan thu 16 (thang 4 1959) ve van de hop tac hoa nong nghiep" (Resolution of the Sixteenth Plenum (April 1959) on Agricultural Cooperativization). Ban chap hanh tung uong Dang lao dong Viet Nam xuat ban.

Dang Phuong Nghi 1969. *Les institutions publiques du Vietnam au XVIIIe siècle*. Paris.

Dang van Ngu. 1974. "Thuc hien nen nep quan ly moi o mot hop tac xa vung dong bang" (Realizing a New Management System in a Delta Cooperative), NCKT 80, 81, 82, and 83.

Dang Phong 1970. *Kinh te thoi nguyen thuy o viet Nam* (The Economy in Primitive Times in Vietnam). Hanoi.

————. 1976. "Ruong cong thoi phong kien o Viet Nam va van de 'phuong thuc san xuat chau A" (Communal Land in Feudal Times in Vietnam and the Problem of the "Asiatic Mode of Production"), NCKT 93, 94.

Dang Viet Thanh. 1962. "Van de mam mong tu ban chu nghia duoi thoi phong kien o Viet Nam—gop y kien voi Nguyen Viet" (The Problem of the "Germs of Capitalism" in Feudal Vietnam—a Comment on Nguyen Viet), NCLS 34, 40.

Dao The An. 1973. "New Rice Strains," *Vietnamese Studies* 38.

Davies, R. W. 1980a. *The Soviet Collective Farm 1929–1930*. London.

————. 1980b. *The Socialist Offensive 1929–1930*. London.

Decoux, J. 1949. *A la barre de l'Indochine: Histoire demon gouvernement générale 1940–45*. Paris.

Despuech, J. 1953. *Le trafic de piastres*. Paris.

Devillers, P. 1952. *Histoire du Vietnam de 1940 a 1952*. Paris.

Dinh Thu Cuc 1977. "Qua trinh tuong buoc cung co va hoan thien quan he san xuat xa hoi chu nghia trong cac hop tac xa san xuat nong nghiep o mien bac nuoc ta" (The Process of Step-by-Step Reinforcement and Improvement of Socialist Production Relations in the Agricultural Producer Cooperatives of the North of Our Country), NCLS 175.

————. 1985. "Nhung buoc dau tien tren con duong di len chu nghia xa hoi cua giai cap nong dan Viet Nam'' (First Steps in the Vietnamese Peasantry's Advance to Socialism), NCLS 4.

Do Duc Hung. 1979. ''Ve tri thuy—thuy loi o nuoc ta nua dau the ky XIX'' (On Flood Control in Our Country During the First Half of the Nineteenth Century), NCLS 18.

Doan Trong Truyen and Pham Thanh Vinh. 1966. *L'édification d'une économie indépendante au Vietnam 1945–1965.* 2d ed. Hanoi.

Donnithorne, A. 1967. *China's Economic System.* New York.

Doumer, P. 1902. *Situation de l'Indochine (1897–1901).* Hanoi.

Duiker, W. J. 1976. *The Rise of Nationalism in Vietnam.* Ithaca.

Dumarest, A. 1935. *La formation des classes sociales en pays annamité.* Lyon.

Dumont, R. 1935. *La culture du riz dans le delta du Tonkin.* Paris.

Duong kinh Quoc. 1938. ''Densité de la population et utilisation du sol en Indochine francaise'' *Congrès international de Géographie.*

―――. 1974. ''Ngan sach cua chinh quyen thuc dan Phap'' (French Colonial Budgets), NCLS 159.

Durand, M. 1952. ''Quelques éléments de l'Univers moral des Vietnamiens,'' BSCI 4.

Duy Hinh. 1963. ''May y kien ve van de phong kien hoa trong lich su Viet Nam'' (Some Problems of Feudalization in Vietnamese History), NCLS 55.

Eastman, L. 1967. *Throne and Mandarins—China's Search for a Policy During the Sino-French Controversy 1880–1885.* Cambridge, Mass.

Elliott, D. W. 1974. ''Political Integration in North Vietnam—The Cooperativisation Period.'' In *Communism in Indochina: New Perspectives*, ed. J. T. Zasloff and M. Brown. Lexington.

―――. 1976. ''Revolutionary Re-Integration: A Comparison of the Foundations of Post-Liberation Political Systems in North Vietnam and China.'' Ph.D. dissertation, Cornell University.

Elvin, M. 1973. *The Pattern of the Chinese Past.* Stanford.

Ennis, T. E. 1936. *French Policy and Developments in Indochina.* Chicago.

Fall, B. B. 1956. *The Viet-Minh Regime.* New York.

―――. 1960. *Le Viet-minh: la république démocratique du Vietnam.* Paris.

―――. 1963. *The Two Vietnams: A Political and Military Analysis.* New York.

Feis, H. 1930. *Europe the World's Banker 1870–1914.* New Haven.

Feyssal, P. 1931. *La réforme foncière en Indochine.* Paris.

Fforde, A. J. 1982. ''Problems of Agricultural Development in North Vietnam,'' Ph.D. Thesis, Cambridge University.

―――. 1983. ''The Historical Background to Agricultural Collectivisation in North Vietnam: The Changing Role of 'Corporate' Economic Power.'' Discussion paper no. 148, Department of Economics, Birkbeck College, London.

―――. 1984. ''Law and Socialist Agricultural Development in Vietnam: The Statute for Agricultural Producer Cooperatives,'' *Review of Socialist Law* 10.

―――. 1984a. ''Macroeconomic Adjustment and Structural Change in a Low-Income Socialist Developing Economy—an Analytical Model.'' Discussion paper no. 163, Department of Economics, Birkbeck College, London.

―――. 1985. ''Economic Aspects of the Soviet-Vietnamese Relationship.'' In *Soviet Interests in the Third World,*'' ed. R. Cassen. London.

―――. 1986. ''The Unimplementability of Policy and the Notion of Law in Vietnamese Communist Thought,'' *South East Asian Journal of Social Studies* 1.

Fforde, A. J., and S. H. Paine. 1987. *The Limits of National Liberation—Problems of Economic Management in the Democratic Republic of Vietnam, with a Statistical Appendix.* London.

Fisher, C. 1964. ''Some Comments on Population Growth in Southeast Asia with Special Reference to the Period Since 1830.'' In *The Economic Development of Southeast Asia,* ed. C. D. Cowan. New York.

Ganiage, J. 1968. *L'expansion coloniale de la France sous la IIIème république 1871-1914*. Paris.

George, A. 1954. *Autonomie économique, cooperation internationale et changements de structure en Indochine*. Paris.

Ginsburgs, G. 1979. "The Genesis of the People's Procuracy in the Democratic Republic of Vietnam," *Review of Socialist Law* 2.

Gluckman, M. 1955. *The Judicial Process Amongst the Barotse of Northern Rhodesia (Zambia)*. Manchester.

————. 1972. *The Idea in Barotse Jurisprudence*. Manchester.

Gordon, A. 1972. "Socialism and Development in North Vietnam," SOAS Left Group.

————. 1981. "North Vietnam's Collectivisation Campaigns: Class Struggle, Production and the 'Middle-Peasant' Problem," *Journal of Contemporary Asia* 1.

Goudal, J. 1938. *Labor Conditions in Indochina*. Geneva.

Gourou, P. 1940. *L'utilisation du sol en Indochine francaise*. Paris.

————. 1955. *The Peasants of the Tonkin Delta*. New Haven.

————. 1975. *Man and Land in the Far East*. Essex.

Gouvernement de l'Indochine. 1931. "Cochinchine," BSEI.

Guillen, P. 1974. "Milieu d'affaires et imperialisme colonial," *Relations Internationales* 1.

Hall, D. G. E. 1961. *Historians of Southeast Asia*. London.

————. 1968. *A History of Southeast Asia*. 3d ed. London.

Hemery, D. 1968. *"L'Indochine de la conquête a la colonisation."* In *L'expansion coloniale de la France sous la IIIème république 1871-1914*, ed. J. Ganiage. Paris.

Henry, Y. 1932. *L'économie agricole de l'indochine*. Hanoi.

Hickey, G. L. 1964. *Village in Vietnam*. New Haven.

Hinton, H. C. 1958. *China's Relations with Burma and Vietnam—a Brief Survey*. New York.

Ho Chi Minh. 1925. *Le procès de la colonisation francaise*. Paris.

————. 1958/1977. *"On Revolutionary Morality."* In *Selected Writings*. Hanoi.

————. 1960/1977. *"The Path Which Led Me to Lenin."* In *Selected Writings*. Hanoi.

Ho Huu Phuoc. 1964a. "Nhan doc bai "may y kien ve van de phong kien hoa lich su Viet Nam (On Reading Some Opinions on Feudalization in Vietnamese History), NCLS 61.

————. 1964b. "Trong lich su Viet Nam den giai doan nao thi ruong dat tu chiem uu the?" (When in Vietnamese History Did Private Land Predominate?), NCLS 69.

Ho Ping-ti. 1952. *Studies in the Population of China 1358-1953*. Cambridge, Mass.

Hoang Anh. 1974. "Réorganization de la production agricole et amélioration d'un pas de la gestion de l'agriculture en direction de la grande production socialiste" Thai Binh Agricultural Conference (August). Photocopy.

Hoang Anh Quoc. 1976. "Ve van de phan phoi va thu nhap cua hop tac xa Giao Tan" (On the Problem of Distribution and Income in Giao Tan Cooperative). UH.

Hoang Kim Giao. 1985. "Ve viec van dung quy luat quan he san xuat phu hop voi tinh chat va trinh do phat trien cua luc luong san xuat trong quan ly kinh te nuoc ta o thoi ky qua do" (On the Use of the Law of Correspondence Between Production Relations and the Quality and Level of Development of Productive Forces in Economic Management During the Transition Period in Our Country), NCKT 5.

Hoang Quoc Viet. 1974. *Tang cuong phap che xa hoi chu nghia trong quan ly kinh te* (Reinforce Socialist Law in Economic Management). Hanoi.

Hoang Uoc. 1968a. "Van de ruong dat o Viet Nam" (The Land Problem in Vietnam). In *Cach mang ruong dat o Viet Nam* (The Land Revolution in Vietnam), ed. Tran Phuong. Hanoi.

————. 1968b. "Nhung cai cach tung phan thuc hien duoi chinh quyen dan chu" (Partial Reforms Realized by the Democratic Administration). In *Cach mang ruong dat o Viet Nam* (The Land Revolution in Vietnam), ed. Tran Phuong. Hanoi.

Hoang van Chi. 1964. *From Colonialism to Communism*. London.

Hodgkin, T. 1981. *Vietnam—The Revolutionary Path*. London.

Houtart, F., and G. Lemercier. 1984. *Hai Van—Life in a Vietnamese Commune*. London.

Huard, P., and M. Durand. 1954. *Connaissance du Vietnam*. Paris.

Humphrey, C. 1983. *Karl Marx Collective—Economy, Society and Religion in a Siberian Collective Farm*. Cambridge.

Huu Hanh. 1980. "Khoan lua" (Rice Contracts), *Tap chi cong san* (Communist Studies) 12.

Huy Vu. 1978. "Vai net ve de dieu, thuy loi o lang xa Viet Nam thoi truoc" (Some Notes on Dykes and Irrigation in the Villages of Old Vietnam), NCLS 180.

Huynh kim Khanh. 1982. *Vietnamese Communism 1925–45*. Ithaca.

Imbert, C. 1885. *Le Tonkin industriel et commercial*. Paris.

Ishikawa, S. 1968. *Economic Development in Asian Perspective*. Tokyo.

Kalecki, M. 1957. "The Problem of Financing Economic Development," *Indian Economic Review* 3.

————. 1972. "Problems of Financing Economic Development in a Mixed Economy." In *Essays on the Economic Growth of the Socialist and the Mixed Economy*. Cambridge.

Karamyshev, V. P. 1959/1961. *Agriculture in the Democratic Republic of Vietnam*. Moscow. (Translated).

Kherian, G. 1937. "Esquisse d'une politique démographique en Indochine," RI 2.

————. 1938. "Les méfaits de la surpopulation deltaique," RI 7/8.

Kornai, J. 1980. *Economics of Shortage*. Amsterdam.

Kleinen, J. 1982. "Regional Development in Vietnam—an Early Communist Rising in Quang-Ngai Province (1930–31)." Working paper no. 5, VZZA, University of Amsterdam.

Lac Ha Long. 1976. "May van de ve quan ly va su dung tu lieu san xuat o hop tac xa Hai Nam" (Some Problems in the Organization and Use of Means of Production in Hai Nam Cooperative), NCKT 93.

Lacan, G. 1944. "Capital and Circulation." In *The Economic Development of French Indochina*, ed. C. Robequain. London.

Lakatos, I. 1970. "Falsification and the Methodology of Scientific Research Programmes." In *Criticism and the Growth of Knowledge*, ed. I. Lakatos and A. Musgrave. Cambridge.

Lam Thanh Liem. 1984. "Collectivisation des terres et crise de l'économie rurale dans le delta du Mekong (1976–1980)" *Annales de Géographie*.

Lavallée, L. 1971. *Problèmes économiques de la République Démocratique du Vietnam*. 2 vols. Paris.

Le Duc Binh. 1968. "Thuc hien cai cach ruong dat" (Realize Land Reform). In *Cach mang ruong dat o Viet Nam* (The Land Revolution in Vietnam), ed. Tran Phuong. Hanoi.

Le Chau. 1966. *Le Vietnam socialiste: Une économie de transition*. Paris.

Le Duan. 1969/1977. "The Vietnamese Revolution—Fundamental Problems, Essential Tasks." In *Selected Writings*. Hanoi.

————. 1974/1977. "The New Stage of Our Revolution and the Tasks of the Trade Unions." In *Selected Writings*. Hanoi.

————. 1975. "Towards a Large-Scale Socialist Agriculture." In *Towards Large-Scale Socialist Agricultural Production*, ed. Le Duan and Pham van Dong. Hanoi.

Le van Hao. 1962. "Introduction a l'ethnologie du dinh." BSEI 1.

————. 1967. "May net ve doi song Viet Nam giua the ky XIX" (Some Notes on Vietnam in the Mid-Nineteenth Century). In *Ky niem 100 nam ngay Phap chiem nam ky* (In Commemoration of the 100th Anniversary of the French Occupation of the South), ed. Truong Ba Can. Saigon.

Le Thanh Khoi. 1955. *Le Viet-nam: Histoire et civilisation.* Paris.

————. 1981. *Histoire du Viet Nam des origines a 1858.* Paris.

Le van Lan. 1977. "Anh huong cua nong thon doi voi cac thanh thi phong kien o Viet Nam" (The Influence of the Countryside on Feudal Towns in Vietnam), NTVNTLS.

Le Kim Ngan. 1977. "Mot so van de ve so huu ruong dat cua lang xa nua dau the hy XIX" (Some Problems of Property in Land in Vietnamese Villages of the Early Nineteenth Century), NTVNTLS.

Le Thanh Nghi. 1981. *Cai tien cong tac khoan mo rong khoan san pham de thuc day san xuat cung co hop tac xa nong nghiep* (Reform Contract Work and Expand Output Contracts to Stimulate Production and Reinforce Agricultural Cooperatives). Hanoi.

Le Nhat Quang. 1976. "Qui mo hop ly cua hop tac xa trong lua o dong bang mien Bac" (The Rational Size of Rice-Growing Cooperatives in the Northern Deltas). NCKT 92.

Le Trong. 1978. "Kinh nghiem ve su phat trien cac hinh thuc to chuc lao dong cua hop tac xa Dinh Cong" (Experiences with the Development of Forms of Production in Dinh Cong), NCKT 101.

————. 1980a. "Kinh nghiem cua cac hop tac xa tien tien va viec nhan dien hinh tien tien trong nong nghiep" (The Experiences of Advanced Cooperatives and the Utilization of Advanced Models in Agriculture), NCKT 113.

————. 1980b. "Ve thu nhap lao dong o hop tac xa nong nghiep" (On Labor Income in Agricultural Cooperatives), NCKT 115.

————. 1980c. "Hiep tac va phan cong lao dong trong hop tac xa nong nghiep" (Cooperation and the division of labor in agricultural cooperatives), NCKT 116.

Lewin, M. 1975. *Political Undercurrents in Soviet Economic Debates.* London.

Lewis, M. D. 1961-62. "One Hundred Million Frenchmen—the Assimilation Theory in French Colonial Policy," *Comparative Studies.*

Limbourg, M. 1956. *L'économie actuelle du Vietnam démocratique.* Hanoi.

Luro, E. 1897. *Le pays d'Annam.* Paris.

————. 1930. *L'Annam d'autrefois.* Paris.

McAleavy, H. 1958. "Varieties of Huong Hoa—a Problem of Vietnamese Law," *SOAS Bulletin.*

————. 1958a. "Dien in China and Vietnam," *Journal of Asian Studies* (May).

————. 1968. *Black Flags in Vietnam.* London.

Popkin, S. 1979. *The Rational Peasant: The Political Economy of Rural Vietnam.* Berkeley.

Rambo, A. T. 1973. "A Comparison of Peasant Systems of Northern and Southern Vietnam." Ph.D. dissertation, Southern Illinois University at Carbondale.

Marr, D. G. 1971. *Vietnamese Anti-Colonialism 1885-1925.* Berkeley.

————. 1981. *Vietnamese Tradition on Trial 1920-1945.* Berkeley.

Marseille, J. 1976. "Les relations commerciales entre la France et son empire colonial de 1880 a 1913," *Relations Internationales.*

Miller, D. B., ed. 1979. *Peasants and Politics: Grass-Roots Reaction to Change in Asia.* London.

Moise, E. E. 1976. "Land Reform and Land Reform Errors in North Vietnam," *Pacific Affairs* (Spring).

————. 1977. "Land Reform in China and North Vietnam—Revolution at the Village Level." Ph.D. dissertation, University of Michigan.

————. *Land Reform in China and North Vietnam.* Chapel Hill.

Morel, J. 1912. *Les concessions de terres au Tonkin.* Paris.

Murray, M. J. 1980. *The Development of Capitalism in Colonial Indochina (1870-1940).* Berkeley.

Mus, P. 1949. "The Role of the Village in Vietnamese Politics," *Pacific Affairs* 22.
————. 1950. *Vietnam; sociologie d'une guerre.* Paris.
Naval Intelligence Division. 1943. *Indochina.* London.
Newberry, D. 1975. "Tenurial Obstacles to Innovation," *Journal Development Studies* (July).
NCLS. 1978. "Ve nong nghiep Viet Nam trong lich su" (On Vietnamese Agricultural History), NCLS 180.
Ng Shui Menh. 1974. *The Population of Indochina.* Singapore.
Ngo Kim Chung. 1975. "Ruong dat tu huu va nhung hinh khai thac ruong dat tu huuo Viet Nam thoi phong kien" (Private Land and Forms of Exploitation of Private Land in Feudal Vietnam), NCLS 85.
Ngo Vinh Long. 1973. *Before the Revolution.* Cambridge, Mass.
————. 1984. "Agrarian Differentiation in the Southern Region of Vietnam," *Journal of Contemporary Asia* 3.
————. 1985. "The National Political Economy of Rural Development in Vietnam, 1975–84," Social Science Research Council Conference on National Policies Towards the Agrarian Sector in Southeast Asia, Chiang Mai, Thailand.
Nguyen The Anh. 1967. "Van de lua o Viet Nam trong ban the ky XIX" (The Rice Problem in Vietnam During the Nineteenth Century), *Su dia* 6.
————. 1971. "Kinh te va xa hoi Viet Nam duoi cac vua trieu Nguyen" (Vietnamese Economy and Society Under the Nguyen). Saigon.
Nguyen Luong Bich. 1968. "Van de so huu ruong dat trong xa hoi phong kien Viet Nam" (The Problem of Landed Property in Feudal Vietnamese Society), NCLS 109.
Nguyen Dong Chi. 1964. "Mot so diem quan he den che do gia dinh cua nguoi Viet Nam thoi co dai" (A Number of Points Relating to the Vietnamese Family System in Ancient Times), NCLS 66.
Nguyen Khac Dam. 1964. "Gop may y kien ve van de ruong tu trong lich su Viet Nam" (Some Opinions on the Problem of Private Land in Vietnamese History), NCLS 65.
————. 1965. "Vai y kien gop cung O. Ho Huu Phuoc ve van de ruong tu" (Some Contributory Opinions for Mr. Ho Huu Phuoc on the Problem of Private Land), NCLS 73.
Nguyen Hoang. 1979. "Hop tac xa Thang Long dua nong nghiep len san xuat lon xa hoi chu nghia" (Thang Long Cooperative Leading Agriculture to Large-Scale Socialist Production), NCKT 108.
Nguyen Manh Huan. 1979. "Hop tac xa nong nghiep Vu Thang" (Agricultural Cooperative Vu Thang), NCKT 107.
————. 1980. "Vu Thang Cooperative, an Advanced Typical Cooperative of Vietnam," UNITAR International Conference on Alternative Strategies for Development and the Future of Asia, New Delhi.
Nguyen Bich Huong. 1976. "Mot so bien phap ve van de day manh nganh trot o hop tac xa Dai Thang" (Some Principles Concerning the Problem of Developing Cultivation in Dai Thang Cooperative). UH.
Nguyen Huy. 1980. "Ve hinh thuc khoan trong lua trong hop tac xa trong lua" (On Forms of Contract in Rice-Growing Cooperatives), NCKT 118.
Nguyen Huu Khang. 1946. *Le commune annamité. Étude historique, juridique et économique.* Paris.
Nguyen van Khoan. 1930. "Essai sur le dinh et le culte du genie tutelaire des villages au Tonkin," BEFEO 30.
Nguyen van Khon. 1969. *Han-Viet tu-dien* (Sino-Vietnamese Dictionary). Saigon.
Nguyen Xuan Lai. 1956. "Étude sur la rente foncière au Vietnam," *Études économiques* 100/101.
————. 1976. "The First Resistance (1945–54)" *Vietnamese Studies* 44.

————, ed. 1977. "Problèmes agricoles (5): La gestion des cooperatives" *Études Vietnamiennes* 51.

Nguyen Lang and Nghiem Phu Ninh. 1972. "To chuc san xuat cong nghiep trong cac hop tac xa nong nghiep" (The Organization of Industrial Production in Agricultural Cooperatives). In *Ve to chuc san xuat trong cong nghiep mien bac nuoc ta* (On Industrial Organization in the North of Our Country), ed. Nguyen Tri. Hanoi.

Nguyen Thieu Lau. 1951. "La mortalité dans Quang-Binh," BEFEO 48.

————. 1951a. "La réforme agraire dans le Binh-Dinh," BEFEO 46.

Nguyen Huu Nghinh and Bui Huy Lo. 1978. "May van de nghien cuu ruong dat cong trong cac lang xa nguoi Viet dau the ke 19" (Some Problems of Research into Communal Land in Vietnamese Villages of the Early Nineteenth Century), *Tap chi dan toc hoc* (Ethnographic Studies) 2.

Nguyen Thanh Nha. 1970. *Tableau économique du Vietnam aux 17e et 18e siècles.* Paris.

Nguyen Nien. 1974. "Quyen quan ly nghiep vu cua xi nghiep doi voi tai san luu dong va cac quy xi nghiep theo che do quan ly moi hien nay," *Luat Hoc* (Legal Studies) 8.

Nguyen van Phong. 1971. *La société vietnamienne de 1882 a 1902 d'après les escrits des auteurs francais.* Paris.

Nguyen Phan Quang and Dang Huy Van. 1966. "Tinh hinh dau tranh giai cap o thoi Gia-Long" (Class Struggle During the Reign of Gia-Long), NCLS 91.

Nguyen Xuan Thang. 1976. "Nghe phu o hop tac xa Giao Tan" (Subsidiary Branches in Giao Tan Cooperative), NCKT 94.

Nguyen Chi Thanh. 1969. *Ve san xuat nong nghiep va hop tac xa nong nghiep* (On Agricultural Production and Agricultural Cooperatives). Hanoi.

Nguyen Xuan Thien. 1976. "Tinh hinh su dung ruong dat o huyen Quynh Luu Nghe Tinh" (Land Use in Quynh Luu District, Nghe Tinh). UH.

Nguyen Khanh Toan. 1980. "Bai noi tai Hoi nghi khoa hoc nghien cuu Nghi quyet lan thu 6 cua Ban chap hanh trung uong Dang" (Speech at the Scientific Conference to Research the Sixth Plenum of the Party Central Committee), NCKT 114.

Nguyen Tri, ed. 1972. *Ve to chuc san xuat trong cong nghiep mien bac nuoc ta* (On Industrial Organization in the North of Our Country). Hanoi.

Nguyen Tran Trong. 1980. "Dang ta va viec dua nong nghiep tu san xuat nho len san xuat lon xa hoi chu nghia o nuoc ta" (Our Party and the Task of Taking Agriculture from Small-Scale Production to Large-Scale Socialist Production), NCKT 113.

Nguyen Khac Vien. 1970. *Experiences vietnammiennes.* Paris.

————. 1974. *Histoire du Vietnam.* Paris.

————. 1979. "Traditional Vietnam—Some Historical Stages," *Vietnamese Studies* 21.

Nguyen Viet. 1962. "Mam mong tu na chu nghia o Viet Nam thoic phong kien" (The Germs of Capitalism in Vietnam), NCLS 35, 36.

O'Ballance, E. 1964. *The Indochina War 1945-54, a Study in Guerrilla Warfare.* London.

Ory, P. 1894. *La commune annamite au Tonkin.* Paris.

Osborne, M. 1969. *The French Presence in Cochinchina—Rule and Response 1859-1905.* Ithaca.

Parish, W. L., and M. K. Whyte. 1978. *Village and Family in Contemporary China.* Chicago.

Pasquier, P. 1907. *L'Annam d'autrefois.* Paris.

Perrot, F. 1902. *La société annamité—la famille, la propriété, l'administration du village et de la province.* Paris.

Pham van Dong. 1975. "Problems of Agricultural Development." In *Towards Large-Scale Socialist Agricultural Production*, ed. Pham van Dong and Le Duan. Hanoi.

Pham Cai Duong. 1965. *Thuc trang cua gioi nong dan duoi thoi Phap thuoc* (The Situation of the Peasantry Under French Rule). Saigon.

Pham Bich Hanh. 1976. "Ve van de phat trien chan nuoi o hop tac xa Dai Thang" (On the Problem of Developing Livestock in Dai Thang Cooperative). UH.

Pham van Kinh. 1971. "Vai y kien ve mot so van de khao co hoc trong quyen 'Kinh te thoi nguyen thuy o Viet Nam" (Some Opinions on a Number of Archaelogical Problems in the Book *The Vietnamese Economy in Primitive Times*), NCLS 136.

————. 1977. "Thu cong nghiep va lang xa Viet Nam" (Artisanal Activity and the Vietnamese Commune). NTVNTLS.

Pham Tran Thinh. 1976. "Su dung hop ly nguon lao dong o hop tac xa Dai Thang" (On the Rational Use of the Labor Supply in Dai Thang Cooperative). UH.

Phan Ke Binh. 1975. *Viet Nam Phong Tuc* (Vietnamese Customs). Paris.

Phan Khanh. 1981. *So thao lich su Thuy loi Viet nam Tap I* (An Outline History of Irrigation and Drainage in Vietnam, vol. 1). Hanoi.

Phan Khoang. 1966. "Luoc su che do xa thon o Viet Nam" (An Outline History of the Communal System in Vietnam) *Van su dia* 1.

Phan Huy Le, ed. 1963. *Lich su che do phong kien Viet Nam* (History of Vietnamese Feudalism). Hanoi.

Pike, D. 1978. *History of Vietnamese Communism 1925-1976*. Stanford.

Population Index. 1945. "French Indochina: Demographic Imbalance and Colonial Policy."

de Pouvoirville, A. 1891. *Le Tonkin actuel 1886-1890*. Paris.

Power, T. F. 1944. *Jules Ferry and the Renaissance of French Imperialism*. New York.

Revue Socialiste. 1953. "Problèmes sociaux du Vietnam."

Robequain, C. 1944. *The Economic Development of French Indochina*. London.

Roberts, S. H. 1929/1963. *The History of French Colonial Policy 1870-1925*. 2d ed. London.

Sainteny, J. 1953. *Histoire d'un paix manquée—Indochine 1945-47*. Paris.

Sansom, R. L. 1971. *The Economics of Insurgency in the Mekong Delta*. Cambridge, Mass.

Scott, J. C. 1976. *The Moral Economy of the Peasant*. New Haven.

Shabad, T. 1958. Economic Developments in North Vietnam, *Pacific Affairs* (March).

Smith, R. B. 1968. *Vietnam and the West*. London.

Smolsky, T. 1937. "Les statistiques de la population indochinoise" Congres Internationale de la Population. Paris.

Spoor, M. 1985. "The Economy of North Vietnam—the First Ten Years: 1955-1964." M.Phil. Thesis, Institute of Social Studies, The Hague.

Statute on Cooperatives. 1970. *Digest of the Statute on Higher Level Agricultural Cooperatives*. U.S. Mission in Vietnam. Saigon (October).

————. 1974. "Statute of Agricultural Cooperatives (High Level)." Rural Publishing House unofficial draft translation. Mimeo.

————. 1979. "Dieu le hop tac xa san xuat nong nghiep bac cao" (Statute for Lower-Level Agricultural Producer Cooperatives). University of Hanoi. Photocopy.

Swain, N. 1981. "The Evolution of Hungary's Agricultural System Since 1967." In *Hungary: A Decade of Economic Reforms*, ed. P. G. Hare, H. K. Radice, and N. Swain. London.

Taboulet, G. 1955-56. *La geste francaise en Indochine*. 2 vols. Paris.

Taylor, J. 1979. *From Modernisation to Modes of Production*. London.

Taylor, K. 1974. "Review of *Lich su Viet Nam* vol. 1," *Journal of Asian Studies* 2.

————. 1976. "The Birth of Vietnam: Sino-Vietnamese Relations to the Tenth Century and the Origins of Vietnamese Nationhood." Ph.D. dissertation, University of Michigan.

Texier, M. 1962. "Le mandarinat au Vietnam au XIXe siècle," BSEI 3.

The Dat. 1981. *Nen Nong nghiep Viet Nam tu sau Cach mang thang tam nam 1945* (Vietnamese Agriculture Since the 1945 August Revolution). Hanoi.

Thich Thien-An. 1975. *Buddhism and Zen in Vietnam*. Los Angeles.

Thompson, V. 1937. *French Indo-China*. London.

To Minh Trong. 1962. "Van de mam mong tu ban chu nghia o Viet Nam" (The Problem of the Germs of Capitalism in Vietnam), NCLS 37.

Toan Anh. 1968. *Nep cu lang xom Viet Nam* (Old Village Customs in Vietnam). Saigon.

Tong cuc thong ke (Statistical Office). 1977. *Tinh hinh phat trien kinh te va van hoa cua mien bac xa hoi chu nghia Viet Nam 1960–1975* (Economic and Cultural Development of Socialist North Vietnam 1960–1975). Hanoi.

—————. 1977a. *Tu dien thong ke* (Statistical Dictionary). Hanoi.

Tran Duc Cuong. 1979. "Nhin lai qua trinh chuyen hop tac xa san xuat nong nghiep tu bac thap len bac o mien Bac nuoc ta" (A Reexamination of the Transition from Lower- to Higher-Level Cooperatives in the North of Our Country), NCLS 187.

Tran Ngoc Dinh. 1970. "Che do so huu ruong dat lon on Nam bo trong thoi de quoc Phap trong tri" (The System of Large-Scale Property in Land in South Vietnam During the Period of French Imperial Rule), NCLS 132.

Tran Kim Ha. 1964 "Ve bai 'may y kien ve van de phong kien hoa trong lich su Viet Nam" (On the Article "Some Opinions on the Problem of Feudalization in Vietnamese History"), NCLS 60.

Tran Quang Lam. 1976. "Ve van de mo rong quy mo cua hop tac xa Giao Tan" (On the Problem of Expansion in Giao Tan Cooperative). UH.

Tran Huy Lieu. 1954. "Van de ruong dat trong lich su Viet Nam" (The Land Question in Vietnamese History), *Tap san su dia, van*, special issue 2.

Tran Phuong, ed. 1968. *Cach mang ruong dat o Viet Nam* (The Land Revolution in Vietnam). Hanoi

Tran Dinh Thien 1976. "May van de ve phan cong lao dong o hop tac xa Van Tien" (Some problems of Labor Allocation in Van Tien Cooperative), UH and NCKT 91.

Tran Nhu Trang. 1972 "The Transformation of the Peasantry in North Vietnam." Ph.D. dissertation, University of Pittsburgh.

Tran Tu. 1970. "Xung quanh cac hinh thuc khai thac ruong lan" (An Overview of Forms of Cultivation of Village Lands), NCLS 133.

Truong Chinh. 1953. *Thuc hien cai cach ruong dat* (Realize Land Reform). Hanoi.

—————. 1959. "Kien quyet dua nong thon mien bac nuoc ta qua con duong hop tac hoa nong nghiep tien len chu nghia xa hoi" (Resolutely Taking Agriculture in the North of Our Country Along the Road of Cooperativization to Advance to Socialism). Speech at the Xth National Assembly, May 20, 1959, Hanoi.

Truong Buu Lam. 1967. *Patterns of Vietnamese Response to Foreign Intervention 1858–1900*. New Haven.

—————. 1968. "L'autorité dans les village vietnamiens au XIXe siècle." In *Leadership and Authority*, ed. G. Wijeyewardene. Singapore.

Truong Huu Quynh. 1977. "Ve nhung quan so huu trong bo phan ruong dat cong o lang xa Viet Nam co truyen" (On Property Relations in Communal Land in Traditional Vietnamese Communes), NTVNTLS.

—————. 1978. "Ve tinh hinh ruong dat va nong nghiep thoi Tay-Son" (On Land and Agriculture During the Tay-Son), NCLS 183.

Truong Nhu Tang. 1986. *Journal of a Vietcong*. London.

Turner, R. F. 1975. *Vietnamese Communism: Its Origins and Development*. Stanford.

Tuan, F. C., and F. W. Crook. 1981. "Planning and Statistical Systems in China's Agriculture." United States Department of Agriculture, Foreign Agricultural Economic Report no. 18.

Ullmer, H. 1937. "Note sur les denombrements des pays d'extrême orients," *Bulletin de l'Institut Internationale de Statistique* 29, 3.

—————. 1937a. "Quelques données démographiques sur les colonies francaises" Con-

grès Internationale de la Population. Paris.
University of Hanoi. N.d. *Giao trinh kinh te nong nghiep* (Course in agricultural economics). Hanoi.
———. 1979. *Giao trinh kinh te cong nghiep* (Course in Industrial Economics). Hanoi.
Uy ban khoa hoc xa hoi Viet Nam (Vietnamese Social Science Commission). 1971. *Lich su Viet Nam*. Vol. 1. Hanoi.
Van Tan. 1970. "Tai sao o Viet Nam tu ban chu nghia khong ra doi trong long che do phong kien?" (Why Did Capitalism Not Arise from Feudalism in Vietnam?), NCLS 130.
Veret, P. 1932. *Au pays d'Annam—les dieux qui meurent*. Paris.
Vassal, G. M. 1910. *On and Off Duty in Annam*. London.
Vickerman, A. 1982. "Collectivisation in the Democratic Republic of Vietnam, 1960–66: A Comment," *Journal of Contemporary Asia* 4.
———. 1984. "Agriculture in the Democratic Republic of Vietnam—the Fate of the Peasantry Under 'Premature Transition to Socialism.'" Ph.D. Thesis, Cambridge University.
Vidal, P. 1887. *Un voyage au Tonkin*. Voiron.
Vien su hoc (Institute of History). 1975. *Viet Nam—nhung su kien 1945–1975* (Vietnam—Dates 1945–1975). Hanoi.
Vietnam Courier. 1978. "A new rice strain."
Vogel, E. F. 1969. *Canton Under Communism*. Cambridge, Mass.
Vo Nguyen Giap and Truong Chinh. 1959/1974. *Van de dan cay* (The Peasant Question). 2d ed. Hanoi (trans. C. P. White).
Vo Nhan Tri. 1967. *Croissance économique de la République Démocratique du Vietnam 1945–1965*. Hanoi.
Vu Tuan Anh. 1974. "Quy mo hop ly cua doi san xuat trong hop tac xa trong lua o dong bang" (The Rational Size of Production Brigades in Rice-Producing Cooperatives), NCKT 81.
Vu thi Dau. 1976. "Van de su dung lao dong nong nghiep o hop tac xa Dai Thang" (The Problem of Agricultural Labor Use in Dai Thang Cooperative). UH.
Vu Trong Khai. 1975. "To chuc san xuat va cai tien quan ly o hop tac xa nong nghiep My Tho" (The Organization of Production and Management Reform in Agricultural Cooperative My Tho), NCKT 88.
Vu Huy Phuc. 1966. "Che do cong dien, cong tho bac ky duoi thoi Phap thong tri" (The System of Communal and Public Lands in North Vietnam Under French Rule), NCLS 87, 88.
Vu Duc Thanh. 1976. "May y kien ve nganh trong trot o Van Tien" (Some Opinions on the Cultivation Branch in Van Tien). UH.
Vu Do Thin. 1954. *Évolution économique du Vietnam*. Paris.
Vu Quoc Thuc. 1951. "L'économie communaliste du Vietnam." Thèse du droit, Paris.
Wadekin, K.-E. 1973. *The Private Sector in Soviet Agriculture*. London.
———. 1982. *Agrarian Policies in Communist Europe*. Dordrecht.
Werner, J. 1984. "Socialist Development: The Political Economy of Agrarian Reform in Vietnam," *Bulletin of Concerned Asian Scholars* 2.
White, C. P. N.d. "Family and Class in the Theory and Practice of Marxism—the Case of Vietnam." Mimeo.
———. 1970. *Land Reform in North Vietnam*. Washington, D.C.
———. 1974. *The Peasant Question* (translation of Vo Nguyen Giap and Truong Chinh, 2d ed., 1959). Ithaca.
———. 1979. "The Peasants and the Party in the Vietnamese Revolution." In *Peasants and Politics: Grass-Roots Reaction to Change in Asia*, ed. D. B. Miller. London.
———. 1981. "Agrarian Reform and National Liberation in the Vietnamese Revolu-

ion: 1920–1957." Ph.D. dissertation, Cornell University.

————. 1982. "Debates in Vietnamese Development Policy", Discussion paper no. 171, Institute of Development Studies, Brighton.

White, J. 1824. *A History of a Voyage to the China Sea.* London.

Whitmore, J. K. 1968. "The Development of Le Government in Fifteenth Century Vietnam." Ph.D. dissertation, Cornell University.

Winch, P. 1963. *The Idea of Social Science and Its Relation to Philosophy.* London.

Woodside, A. B. 1970. "Decolonization and Agricultural Reform in North Vietnam," *Asian Survey* (August).

————. 1971. *Vietnam and the Chinese Model: A Comparative Study of Nguyen and Ching Civil Government in the First Half of the Nineteenth Century.* Cambridge, Mass.

————. 1976. *Community and Revolution in Modern Vietnam.* Boston.

Yu I. 1978. "Law and Family in 17th and 18th Century Vietnam." Ph.D. dissertation, University of Michigan.

Zelinsky, W. 1949–50. "The Indochinese Peninsula—a Demographic Anomaly," *Far Eastern Quarterly.*

Index

Accumulation fund, 37, 123-27
Active cooperative, 50–51, 54–55, 80, 84, 90, 92, 127, 136, 141. *See also* Dai Thang cooperative (Quynh Luu); Dai Thang cooperative (Xuan Thuy); Giao Tan cooperative; Hai Nam cooperative; My Tho cooperative; Thang Long cooperative; Van Tien cooperative
Administrative units. *See* Cooperative; Commune; District; Family; Hamlet; Province; Traditional rural organization; Village
Agrarian Question: context of study of, 3–8; early 1970s and, 81; historical background, 7–13, 185–88; solution to, 194–211
Agricultural input, 19, 20, 121–22
Agricultural output, 18–21, 121–27, 144
Agricultural producer cooperative. *See* Cooperative
Amalgamation, 12–13, 62–63, 77, 84, 86, 89, 90, 92–94, 97, 170, 203, 208, 210; of Dinh Cong, 62; of Giao Tan, 171; of My Tho, 98–99, 105–106; of Thang Long, 172; of Vu Thang, 62. *See also* New Management System
Artisanal production. *See* Subsidiary branch
Assembly. *See* General Assembly of the cooperative

Authority, 48–50, 114, 115
Azolla, 165–66

Bonuses and penalties, 122–27, 177, 204. *See also* III-point contract system; Workpoint system
Branches of rural production. *See* Cultivation branch; Livestock branch; Subsidiary branch
Bricks and tiles, manufacture of. *See* Subsidiary branch
Brigade/team, 29–30, 33–34, 35, 39, 90, 127, 153–56, 161, 184–87, 193; active cooperative and, 54; in Dai Thang (Quynh Luu), 153; in Dai Thang (Xuan Thuy), 153, 156, 163, 177; in Dinh Cong, 64–68, 153, 156, 163; distribution systems and, 122, 123–25; in Giao Tan, 123–25, 163, 165; in Hai Nam, 124, 163; leadership of, 34, 42, 104, 111–12, 113, 118; model cooperative and, 52–53, 55; in My Tho, 98–99, 101–14, 156; NMS centralization and, 40–41, 43–44, 49, 52, 60, 83–86, 90, 105–14, 116, 180; nominal cooperative and, 51, 53–54, 98–114, 127, 176; output contracting system and, 203–4; pig rearing and, 134, 136; rice producing, 55, 63, 65, 85–86, 106, 120, 127, 152, 158–59, 163, 169, 206; specialized, 60,